Searching for the Seventies

The DOCUMERICA Photography Project

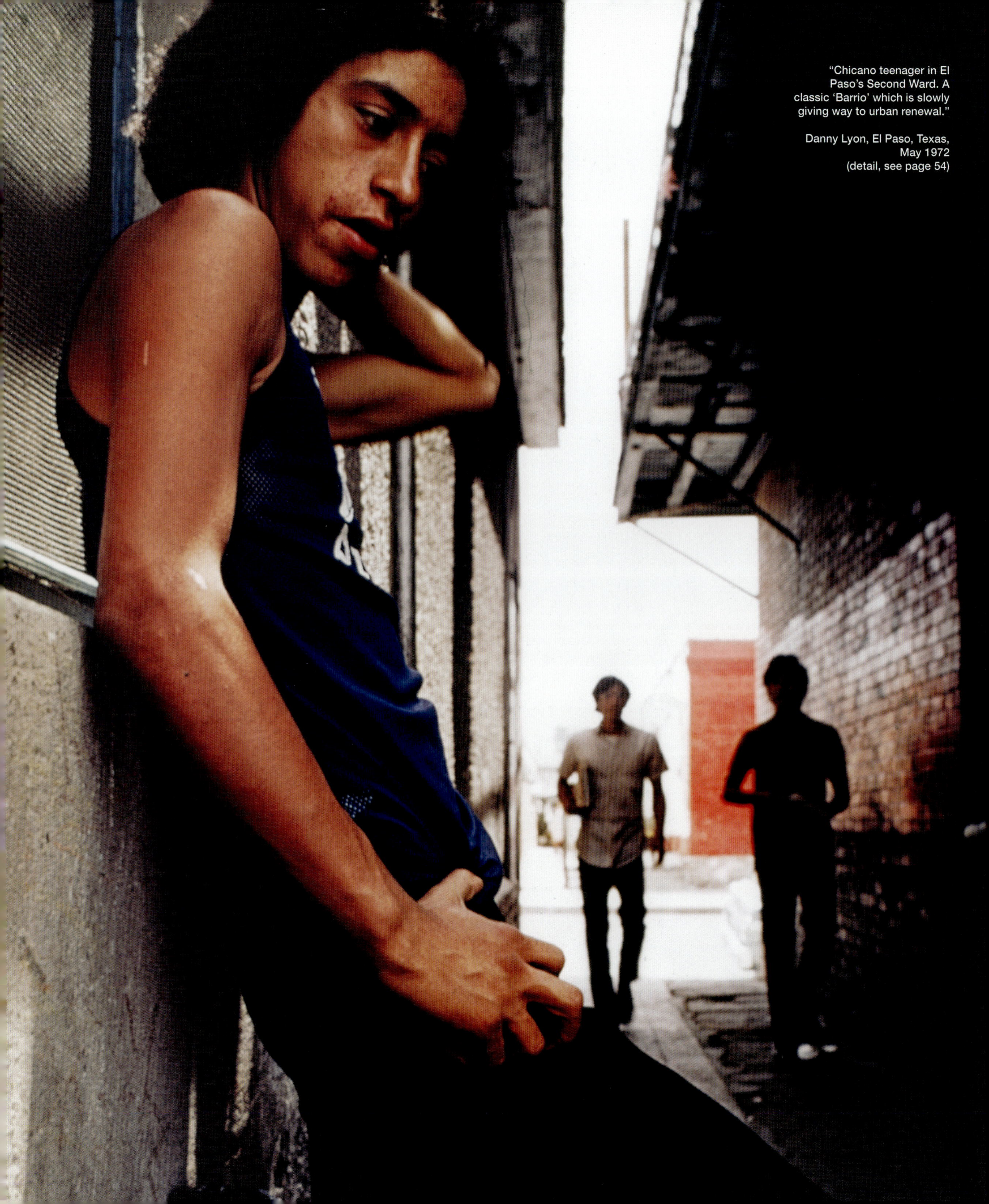

"Chicano teenager in El Paso's Second Ward. A classic 'Barrio' which is slowly giving way to urban renewal."

Danny Lyon, El Paso, Texas, May 1972
(detail, see page 54)

Searching for the Seventies

The DOCUMERICA Photography Project

By

Bruce I. Bustard

With a message from
David S. Ferriero
Archivist of the United States

Foreword by **William D. Ruckelshaus**

The Foundation for the National Archives, Washington, DC
in association with D Giles Limited, London

Searching for the Seventies
The DOCUMERICA Photography Project

This book is based on the exhibition "Searching for the Seventies: The DOCUMERICA Photography Project," presented in the Lawrence F. O'Brien Gallery at the National Archives Building, Washington, DC, from March 8, 2013, to September 8, 2013.

Written by Bruce I. Bustard
With a message from David S. Ferriero, Archivist of the United States
Foreword by William D. Ruckelshaus, First and Fifth Administrator, U.S. Environmental Protection Agency
Designed by Amanda Perez
Edited by Patty Reinert Mason and Benjamin Guterman

First published in 2013 by GILES
an imprint of D Giles Limited
4 Crescent Stables, 139 Upper Richmond Road
London, SW15 2TN, UK
www.gilesltd.com

ISBN (hardcover): 978-1-907804-15-1
ISBN (softcover): 978-0-9841033-7-9

Library of Congress Control Number: 2012948300

Front cover:
"Hitchhiker with his dog, 'Tripper,' on U.S. 66. U.S. 66 crosses the Colorado River at Topock."
Charles O'Rear, Yuma County, Arizona, May 1972 (detail, see page 52)

Back cover:
"Abandoned automobiles and other debris clutter an acid water and oil filled five acre pond. It was cleaned up under EPA supervision to prevent possible contamination of Great Salt Lake and a wildlife refuge nearby."
Bruce McAllister, near Ogden, Utah, April 1974 (detail, see page 14)

End papers:
"California Western Railroad"
Tomas Sennett, Northern California, May 1974 (412-DA-404)

Unless otherwise noted, all photographs in this book are digital prints made from DOCUMERICA slides held by the National Archives and Records Administration and are among Records of the Environmental Protection Agency, Record Group 412. Minor flaws on some of the photographs reflect the condition of the original slides when they came to the National Archives. Captions in quotation marks are as written by the project's photographers, with minor editorial corrections. Photograph identification numbers (e.g., 412-DA-9490) appear at the end of each caption.

Printed and bound in China.

Contents

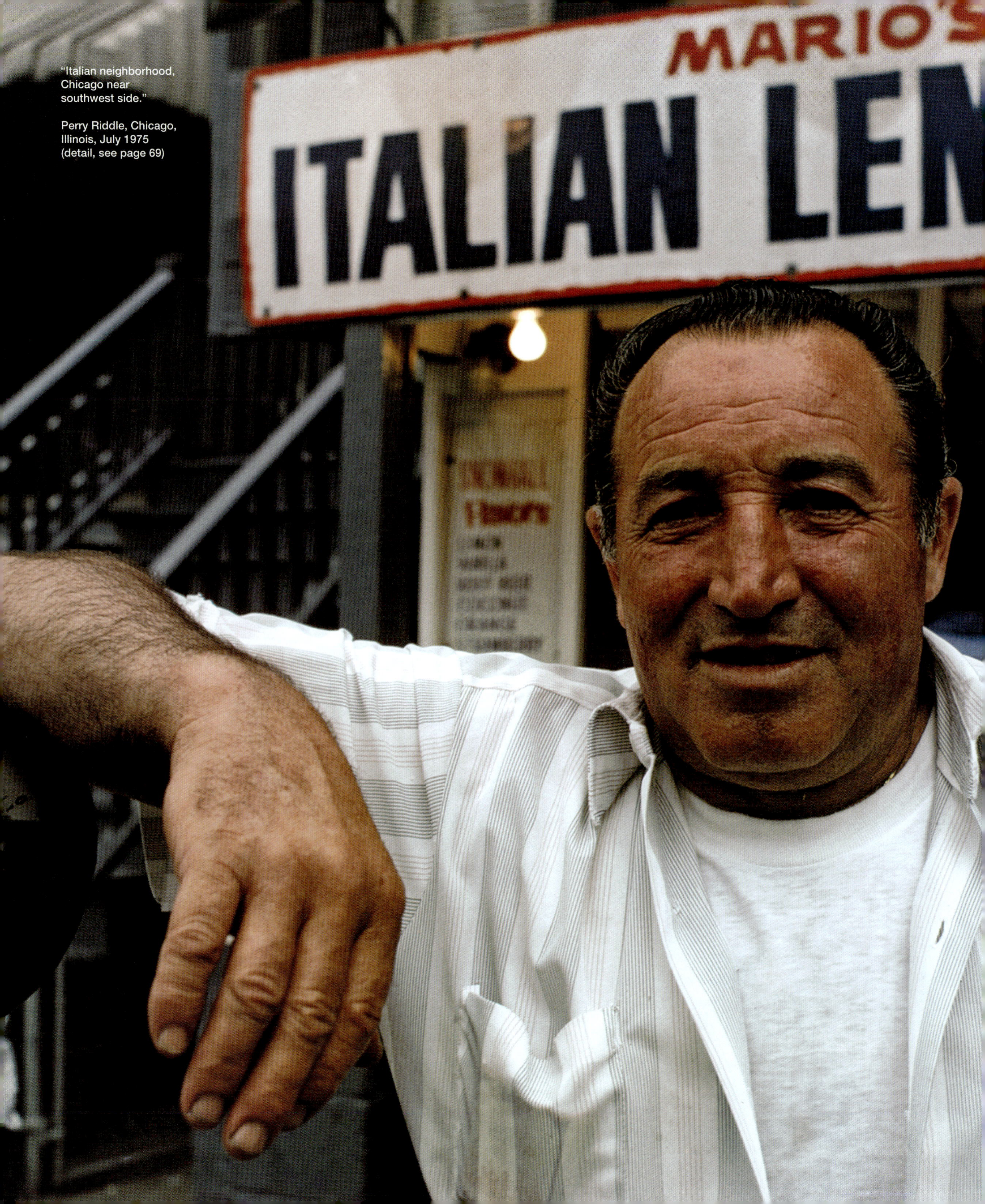

"Italian neighborhood, Chicago near southwest side."

Perry Riddle, Chicago, Illinois, July 1975 (detail, see page 69)

Message from the Archivist of the United States

The National Archives is best known as the home of the original Declaration of Independence, the U.S. Constitution, and the Bill of Rights. But these treasured documents—precious as they are—represent but three of the *billions* of Federal Government records we hold in trust for the American people.

We preserve 2–3 percent of all records created by the United States Government—those that are important for legal or historical reasons. That may not sound like very much, but we now have approximately 12 billion pages of records. That's 1.4 million trees—enough paper that if the pages were laid end-to-end, it would circle the globe 87 times! We hold nearly 500 terabytes of electronic records, and miles of film and video records.

The National Archives is also one of the world's great repositories of historical photographs, with 40 million in our care. This ever-growing number of prints, negatives, slides, transparencies, and digital images are held at the National Archives in College Park, Maryland, as well as in Presidential libraries and museums and regional facilities around the nation.

Among them are glass plate negatives from the Civil War taken by Mathew Brady's studio; albumen prints of exploration of the American West by Carleton Watkins, Timothy O'Sullivan, and William Henry Jackson; and 20th-century photographic prints by Dorothea Lange, Ansel Adams, and Lewis Hine. The stacks of the National Archives also contain mug shots from Federal prisons, passport photos of immigrants, and yearbooks from high schools in World War II Japanese-American internment camps. We also safeguard aerial photographs surveying the Kremlin, Pearl Harbor, the Warsaw Ghetto, and Hiroshima—before and after the atomic bombing.

The National Archives Experience exhibition "Searching for the Seventies: The DOCUMERICA Photography Project" and this accompanying publication highlight a group of photographs from our more recent past. These photos, created by the Environmental Protection Agency's Project DOCUMERICA, include the expected images of smog, polluted rivers, and waste dumps. But the photos also capture the decade's fashions, issues, and lifestyles—a fascinating time capsule of 1970s America.

I hope these photographs will inspire you to visit the National Archives to explore more of our photographic holdings. Meantime, we continue to gather digital photographs from many Federal agencies, putting hundreds of thousands of them online.

These images are part of our nation's history, and we are proud of our role in preserving them for future generations.

David S. Ferriero

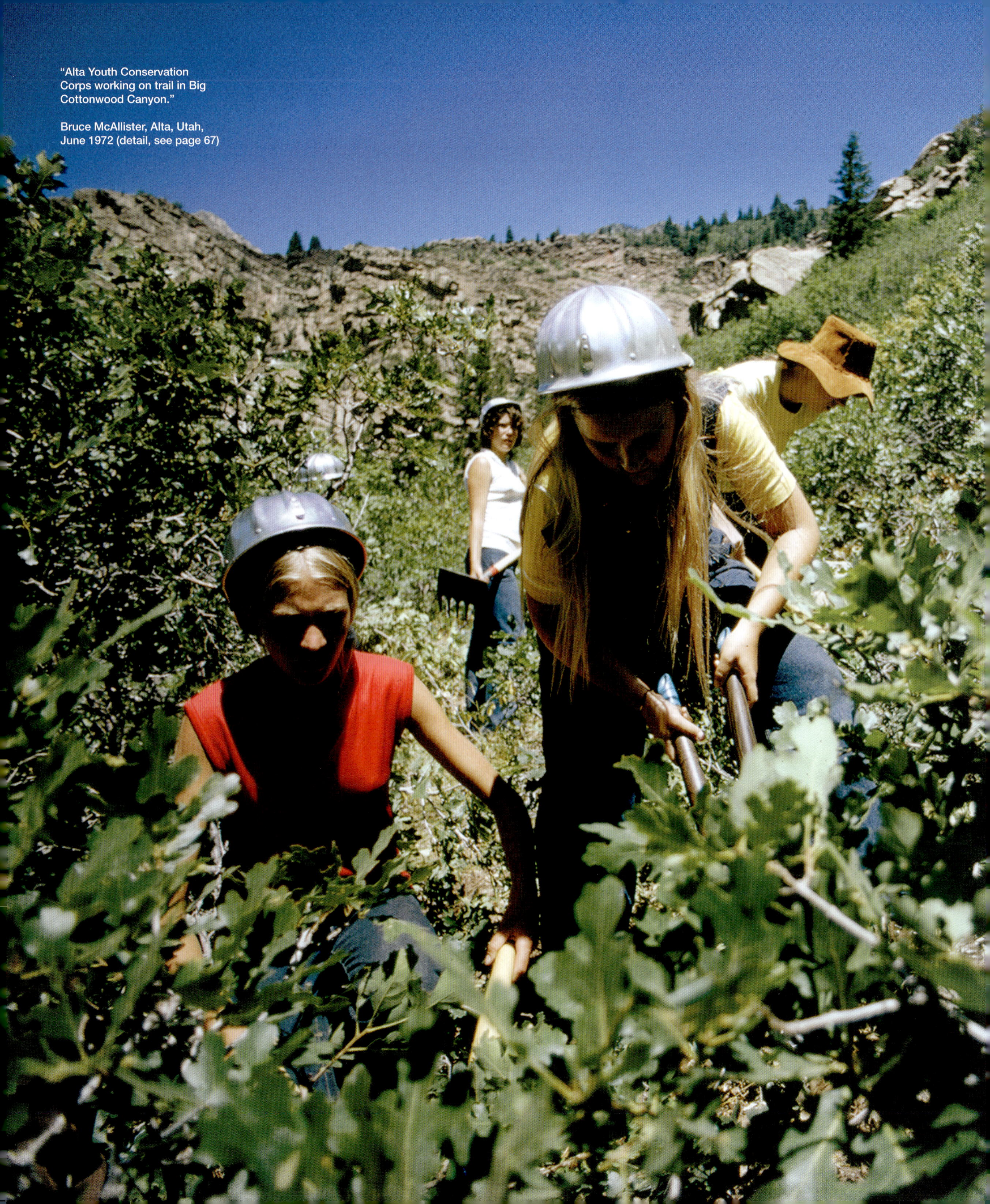

"Alta Youth Conservation Corps working on trail in Big Cottonwood Canyon."

Bruce McAllister, Alta, Utah, June 1972 (detail, see page 67)

Foreword

The decade of the 1970s was a time when the public demanded that we as a society do something about pollution. And it was gross pollution—pollution we could see, taste, smell, and sometimes feel—that sparked that outpouring of concern and demand.

Photographs by Ansel Adams, among others, were iconic symbols of the great American landscape and spurred a public increasingly angry and frustrated by what they saw as the desecration of that landscape and harm to its creatures.

I also believe the impact of color television on the explosion of environmental awareness and concern has never been fully appreciated. From 1964, soon after Rachel Carson's *Silent Spring* was published, to 1970, the year of the first Earth Day, the percentage of houses with color television grew over 1,000 percent, and by 1972, over half of the homes in our country had color television. Americans could now see yellow outfalls into blue waters, and air so dirty it obscured the horizon. Those images defined a desecration and a threat to what people regarded then as "the environment" that no newspaper story, no speech, no black-and-white photo ever could.

That public pressure resulted in the creation of the Environmental Protection Agency (EPA) and the passage of seminal laws that remain today the fundamental framework of our environmental protection system.

It is hard to convey what it was like then: the level of excitement, enthusiasm, and expectation that greeted the early efforts of our government to address the obvious pollution problems . . . problems that created the public demand for that action in the first place. The EPA was besieged—and that is not too strong a word—with thousands of applications from people all over the country who wanted to come to work there, to be involved in what they saw as a chance to make a difference.

Soon after the EPA was created, I met with Gifford Hampshire, the man behind DOCUMERICA, and his boss, Tom Hart, who had just joined the Agency as head of public affairs. The DOCUMERICA idea that Hampshire proposed, and that Hart enthusiastically supported, was extremely ambitious. It was really far broader in scope and less structured in its format than Roy Stryker's Farm Security Administration photography project—the intellectual and artistic foundation for DOCUMERICA.

Pictures of pollution can be powerful, but they can also lose their power to engage us and to prod us to action if that is all you see. What DOCUMERICA tried to do, and what Giff Hampshire and I spoke about when he asked for approval of the project, was to find ways not only to capture the spirit of concern for the environment that was in full voice in this country, but also to show how broad was the concept of the environment . . . that it was more than beautiful landscapes or flaming rivers; it also involved real people and the places where they lived.

The physical environment affects us all but in many different ways. Gross pollution offends our senses . . . even if it was, as it used to be called, "the smell of money." But, as we know now, the invisible pollution in our air and water—the toxins and other substances that are impossible to detect with our basic senses—impact our health and thus our future.

So while DOCUMERICA did capture scenes of environmental harm and defilement, its focus on people and the environments in which they lived and worked anticipated, in an almost uncanny way, what is today the primary focus of the EPA: the protection of public health.

While many of these pictures may seem dated now, they serve as a powerful reminder that protecting our environment is a job that is never really over. It's not like building a bridge or paving a highway—it is something that we must keep everlastingly at. That message may be the most important and abiding legacy of DOCUMERICA.

William D. Ruckelshaus
First and Fifth Administrator, U.S. Environmental Protection Agency

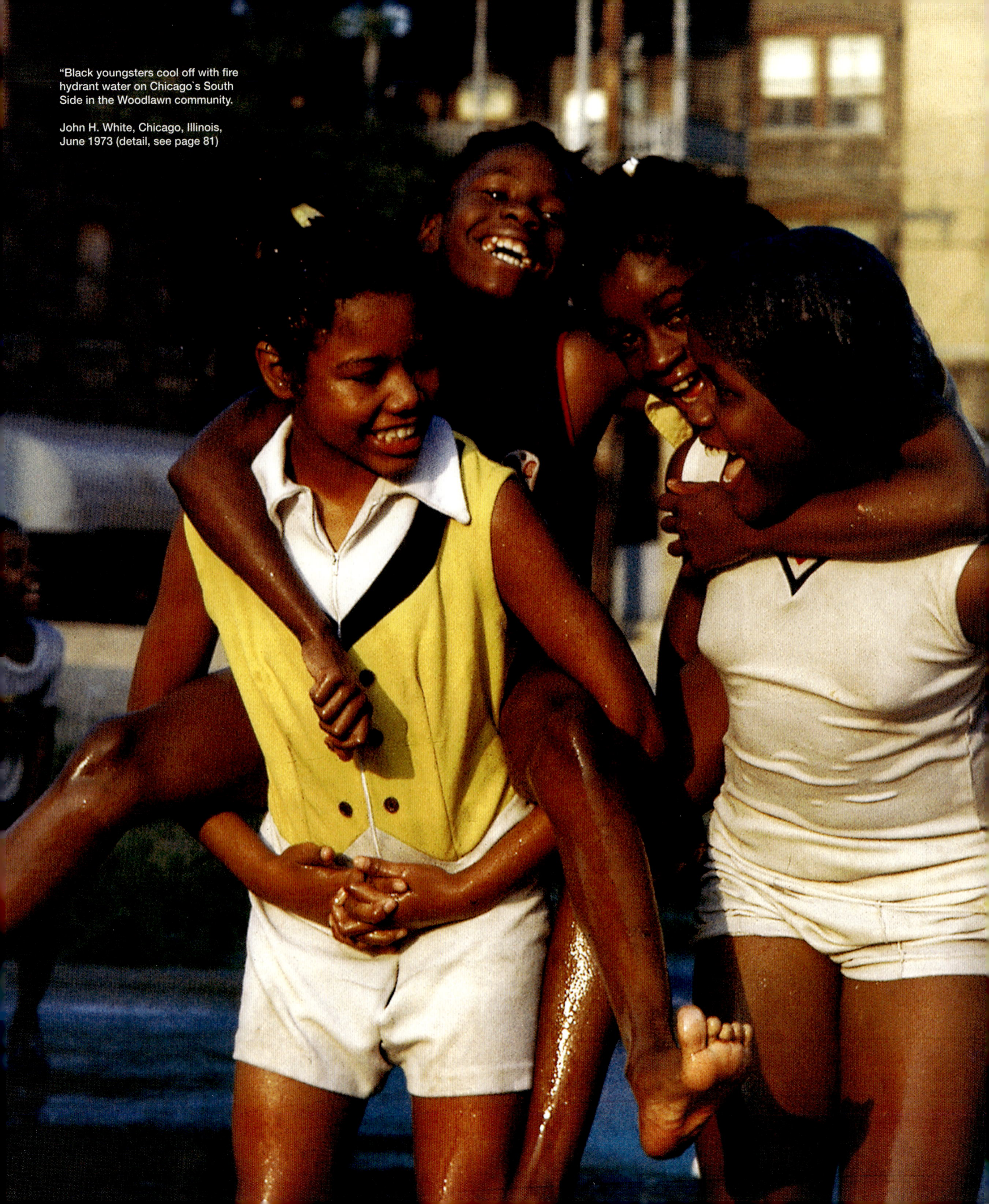

"Black youngsters cool off with fire hydrant water on Chicago`s South Side in the Woodlawn community.

John H. White, Chicago, Illinois, June 1973 (detail, see page 81)

Acknowledgments

A Beatles song reminds me that "I get by with a little help from my friends." Nowhere was this truer than with the "Searching for the Seventies: The DOCUMERICA Photography Project" exhibition and book. The exhibition was created at the National Archives in Washington, DC, in the Lawrence F. O'Brien Gallery under the able leadership of James Gardner, Executive for Legislative Archives, Presidential Libraries, and Museum Services. Christina Rudy Smith, Director of the National Archives Experience, supported the idea of a DOCUMERICA exhibition for many years when it seemed there was little chance it would become a reality and offered many helpful editorial suggestions. Daniel Falk and Amanda Perez took ideas and words and made them real with their wonderfully evocative exhibition and catalog designs. James Zeender, Karen Hibbitt, and Alexis Hill served as exhibition registrars and oversaw retrieving documents, arranging for and managing scans of photographs, and preparing the exhibition to travel. Patty Reinert Mason and Benjamin Guterman edited the exhibition catalog and saved me from numerous embarrassing misspellings, clunky phrases, and wrong word choices. Ben also edited the exhibition text.

In addition, I am also grateful to the following National Archives staff for their assistance: Hannah Bergman, Terry Boone, Laura Diachenko, Cecilia Epstein, Michelle Farnsworth, Jill James, Miriam Kleiman, Mary Krackowiak, Edward McCarter, Holly Reed, Jeffery Reed, Dan Rooney, and Jennifer Seitz. My colleagues in the Center for the National Archives Experience—Katherine Chin, Catherine Farmer, Michael Hussey, Michael Jackson, Jennifer Johnson, Alice Kamps, Patrick Kepley, Darlene McClurkin, Doug Mcrae, Ray Ruskin, and Will Sandoval, Annie Wilker, Morgan Zinsmeister—were generous with their time and talents.

Several colleagues from both inside and outside the Archives provided their expertise. At the Archives, Nicholas Natanson and Jerry Simmons shared their knowledge of DOCUMERICA and greatly improved the exhibit script and catalog essays. Professor Bruce J. Schulman of Boston University added his vast knowledge of the 1970s to the project. Barbara Shubinski of the Rockefeller Archives reviewed the exhibit script and catalog and inspired me with her wonderful dissertation on DOCUMERICA. Jeanethe Falvey of the Environmental Protection Agency spread the word about the exhibition and helped me make connections between DOCUMERICA and the EPA's State of the Environment photography project.

The Foundation for the National Archives, under the leadership of President A'Lelia Bundles and the direction of Thora Colot, generously supported the publication of this catalog and the exhibition's marketing. Director of Publications Patty Reinert Mason managed the catalog project and edited the book. Kathleen Lietzau provided administrative and logistical support. Franck Cordes, the Foundation's Director of Administration and Marketing, was an enthusiastic advocate for the exhibition and led the marketing effort and Director of Development Stefanie Mathew led the fundraising effort. I also thank Foundation Board members Lawrence F. O'Brien III and Nancy Folger for their support.

Finally, I am indebted to Archivist of the United States David S. Ferriero and former EPA Administrator William D. Ruckelshaus for their wonderful contributions to this book.

Exhibitions and publications are enormously collaborative undertakings. I am indebted to everyone who has been a part of "Searching for the Seventies."

Bruce I. Bustard

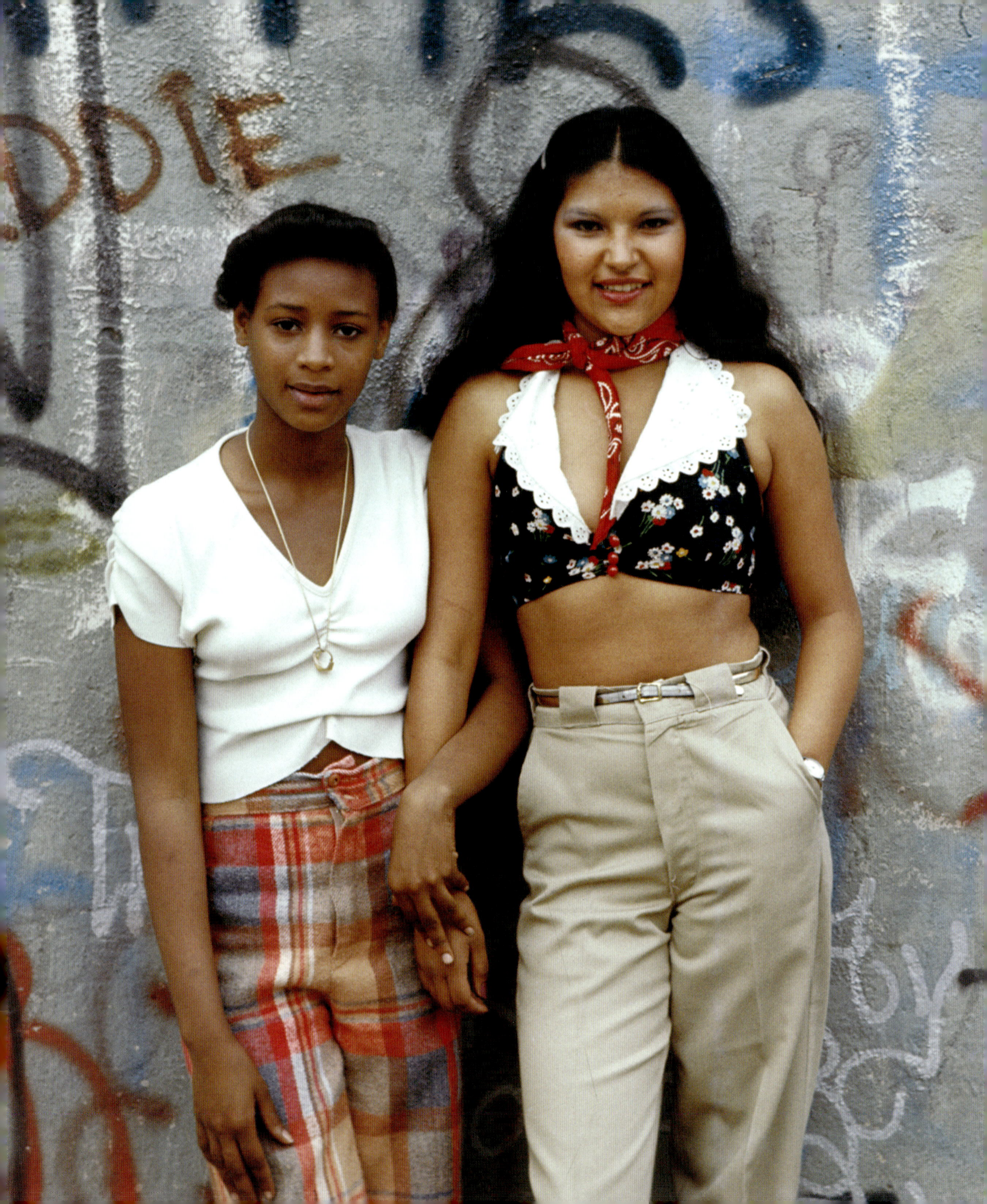

Introduction

Searching for the Seventies: The DOCUMERICA Photography Project

Bad fashion, odd fads, and disco dance music sum up the 1970s for many Americans. We contrast those years to the politically committed 1960s civil rights triumphs, antiwar protests, Great Society reforms, and New Left radicalism. They seem a poor relation to the economically booming 1980s Republican resurgence, yuppies, tax cuts, and leveraged buyouts. The decade of the seventies is remembered as a time of soaring inflation, political corruption, and loss of prestige around the world. It was an era of political and economic drift—the self-indulgent "me decade."

But the 1970s were much more than leisure suits, streaking, and disco. Their importance goes beyond high gas prices, Watergate, and Vietnam. During the seventies, profound changes took root in our politics, society, and economy. Many Americans lost confidence in their political leaders and began to doubt government could solve their problems. We pushed the boundaries of personal freedom by challenging accepted beliefs about gender roles, sexuality, dress, marriage, the family, race, and age. Building on the successes of the African American freedom movement, other groups—including Hispanics, Native Americans, gays and lesbians, seniors, and the disabled—asserted their rights. Women, especially, used recently passed laws and court decisions to enter the workforce in greater numbers, gain educational equality, and take more control over their sexual and reproductive lives.

The 1970s also saw huge changes to the economy as Americans lost manufacturing jobs, endured two oil embargoes, and suffered from "stagflation"—an unprecedented combination of slow growth and high inflation. Traditional high-paying manufacturing jobs began to vanish, especially in the North and Midwest, as businesses "outsourced" jobs to the "sunbelt" in the South and West as well as overseas. Service-sector jobs in this "deindustrialized" economy would be lower-paying and nonunion. As workers' wages fell behind the rising cost of living, poverty rates, which had reached their lowest point in 1973, began to rise. Many citizens came to believe that the best the Federal Government could do was to get out of the way of free market solutions; others began to think that the United States was facing a new era of economic limits, expensive energy, and lowered expectations.[1] Such transformations refute the conventional wisdom that the years of the 1970s were

Opposite: "Two Latin girls pose in front of a wall of graffiti in Lynch Park in Brooklyn, New York City."

Danny Lyon, Brooklyn, New York, June 1974 (detail, see page 48)

inconsequential and forgettable. *Searching for the Seventies: The DOCUMERICA Photography Project* invites you to take a new look at those years through the lens of a Federal Government photography project called Project DOCUMERICA. Created by the Environmental Protection Agency (EPA) in 1971, DOCUMERICA was born out of the decade's environmental awakening, producing striking photographs of that era's environmental problems and achievements. As you might expect, the photographers hired by the EPA took thousands of photographs depicting pollution, waste, and blight, but they were given the freedom to capture the era's trends, fashions, and cultural shifts. The result is an amazing archive and a fascinating portrait of America from 1972 to 1977.

DOCUMERICA logo, 1971
National Archives, Records of the Environmental Protection Agency

DOCUMERICA was the brainchild of Gifford Hampshire. Born in Kansas in 1924, "Giff" Hampshire grew up during the Great Depression. His father was an amateur photographer who let his son watch as images magically appeared in trays of developing solution. After his father lost his job in the mid-1930s, Hampshire lived on an uncle's farm where he witnessed Dust Bowl scenes similar to those captured by photographers working for the U.S. Government's Historical Section of the Resettlement Administration/Farm Security Administration (FSA), directed by Roy Stryker. Those two influences came together when Hampshire returned home from the Army Air Corps after World War II and entered the University of Missouri. There, he took journalism courses with a professor who brought former FSA photographers Russell Lee and Esther Bubley to campus. Decades later, he recalled being "terribly excited" hearing Lee and Bubley talk about their experiences, and he pointed to those conversations as inspiring a hope to someday "do something . . . comparable to the FSA and Roy Stryker."[2]

After graduating from Missouri in 1949, Hampshire took a job with the Fairchild Camera Corporation, promoting the company's new photoengraving machine. In 1954, he left Fairchild to become a photo editor with *National Geographic*. Then, in 1961, after working briefly for the United States Information Agency, he moved to the Food and Drug Administration (FDA), where he spent the next 10 years as a magazine editor and speechwriter. During his years with the FDA, he occasionally promoted the idea of a new documentary photography project based on the FSA, but with no success.[3]

In 1970, Hampshire's chances of fulfilling his dream improved dramatically. Beginning in the early 1960s, environmentalism, a critique of American society aimed at protecting and improving the nation's natural environment, emerged from sources as diverse as scientists, White House policymakers, middle-class women reformers, countercultural activists, and New Left radicals. By 1970, environmentalism was a powerful social movement and a potent political force. Journalists covered the movement extensively, authors wrote best-selling books warning of the dire consequences of environmental threats, and politicians of all persuasions appealed to voters with ecologically friendly ideas. High-profile environmental disasters such as the January 1969 Santa Barbara, California, oil spill and the June 1969 fire on the Cuyahoga River in Cleveland, Ohio, gave substance to such worries. On April 22, 1970, this environmentalism coalesced when millions of Americans celebrated the first Earth Day.[4]

The White House and Congress reacted to those heightened environmental concerns with executive and legislative action. President Richard Nixon thought that the environmental movement would be short-lived and that its ranks were filled with radicals, but his political instincts told him the movement was important to his re-election. He established an Environmental Quality Council within the White House, sent Congress a 37-point message on the environment, and used his 1970 State of the Union to proclaim a "decade of environmental transformation." Congress passed a tough new Clean Air Act and the National Environmental Policy Act. When Nixon signed the latter into law on January 1, 1970, it set up the EPA and gave it strong regulatory powers. In November, Nixon named William Ruckelshaus its first Administrator.[5]

If the EPA's creation was a milestone in American environmental history, it also changed Giff Hampshire's career. In late 1970, he transferred from the FDA to the EPA and became Deputy Director of its Office of Public Affairs. Still fascinated with the FSA, he approached aides to Ruckelshaus, raising the notion of a photography project dealing with environmental issues. The young staff members were intrigued and urged him to go to Ruckelshaus. The Administrator, perhaps looking for ways in which the new agency could make its mark, was enthusiastic, and asked Hampshire to head the undertaking, which was given the name Project DOCUMERICA.

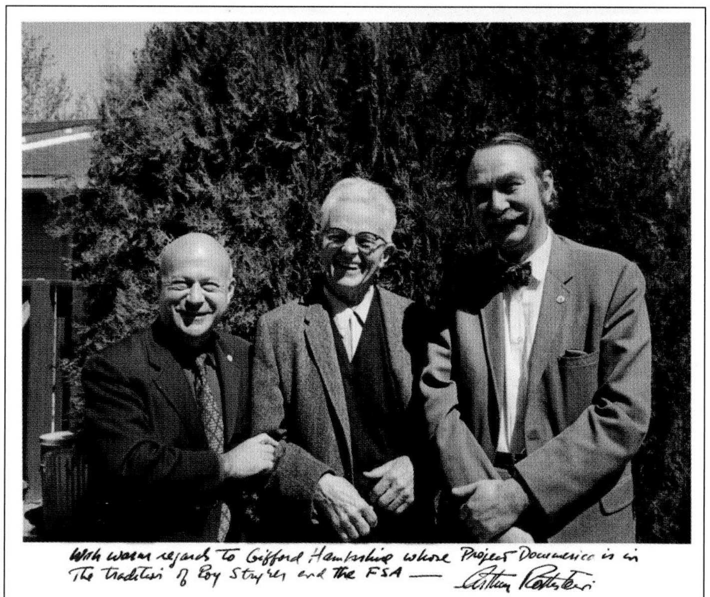

Left to right: Arthur Rothstein, Roy Stryker, and Gifford Hampshire, 1972
Courtesy of the Hampshire family

In November 1971, Ruckelshaus and Hampshire introduced DOCUMERICA to the public and the media. The project would initially employ about 50 photographers who would "pictorially document the environmental movement in America in this decade." Its photographs would establish "a visual baseline" of 1970s environmental images from which improvements could be measured. For example, DOCUMERICA photographers might "record . . . current air pollution problems" and later re-photograph the same locations to chart progress. Other subjects would include water pollution, solid waste management, pesticide control, noise pollution, and radiation reduction. The photographs would demonstrate the "impact of the environmental problem," tally the social and economic costs of environmental change, document "the environmental movement itself," and, more broadly, "depict Americans 'doing their environmental thing.'"[6]

But Hampshire wanted even more from DOCUMERICA. He admitted that many assignments such as sewage treatment plants would prove to be "dull visual material." He wanted to capture what he called "the human condition" and how it related to environmental issues. Taking environmentalist Barry Commoner's "First Law of Ecology" as the project's credo, Hampshire told his photographers that "Everything is connected to everything else." If DOCUMERICA was to be more than "a dismal record of environmental problems and what EPA and others do to solve them," its photographs needed to show "what the people of the nation do, or are, or think, or feel, or whatever it is that's 'connected to everything else.'" Such an expansion of the project's mission was crucial because the environmental movement was "changing lifestyles" and even "closing the generation gap with the emergence of a new environmental ethic."[7]

From the start, DOCUMERICA drew on the FSA for inspiration. Hampshire saw his project as "a contemporary application of the photojournalistic principles established by the FSA." He held up the Depression-era agency as its model in the initial press release about DOCUMERICA, and he hired former FSA photographer Arthur Rothstein as an advisor. When photographers asked for guidance he pointed them to F. Jack Hurley's *Portrait of a Decade: Roy Stryker and the Development of Documentary Photography in the Thirties*, a history of the FSA. While there were differences, such as DOCUMERICA's reliance on color photography and its larger staff, as a whole the project had a similar nationwide scope and its photographers explored comparable issues such as change and continuity, the nature of progress, and American values through in-depth photographic stories. Consciously or not, DOCUMERICA photographers often took photographs that seem inspired by (or even imitative of) documentary techniques pioneered by the FSA in the 1930s.

But, as Hampshire pointed out, he was "no Roy Stryker." He hoped to avoid the "propaganda" for agency programs he noticed in many FSA photos, and he emphasized that his photographers would have the freedom they cherished to pursue the photographs they intended. Unlike Stryker, Hampshire did not send out lists of images he needed for the files, nor did he offer much in the way of guidance or critique. "I'm not saying," he wrote to a photographer, "that I cannot direct photographers into meaningful assignments . . . I'm saying that I should not. I contend the best photography comes out of the photographer's intellectual involvement as well as emotional." He bragged that DOCUMERICA was "a photographer's project."[8]

Also unlike the FSA, DOCUMERICA did not hire a permanent photographic staff. Instead, it contracted out assignments to local photographers who were paid $150 a day plus film and expenses for brief assignments. Typically, these photographers proposed ideas after hearing from EPA experts who spoke to them about the environmental issues of their region. Hampshire and Rothstein would then meet with them and ask, "Well, what did you hear that set you off, as a subject?" As they completed their assignments, photographers sent their exposed film to Washington, DC. It was processed in a New York City lab, and then the slides were shipped to the photographers, who selected and captioned the slides they wanted to submit. Those slides went back to Washington, DC, where Hampshire and his staff added them to the file. Photographers received full credit for their shots, and any rejected images were theirs. Accepted slides became Government property and were in the public domain.

Over the next six years, Hampshire would hire about 70 photographers who would complete 115 assignments covering all 50 states. Assignments varied greatly. They could be as narrow as Gene Daniels's effort to photograph "fly fishermen for conservation" or as general as Frank Aleksandrowicz's to record "the environmental problems" of metropolitan Cleveland, Ohio. A few assignments were directly tied to the EPA's mission and needs. Bruce McAllister, for example, photographed the EPA's cleanup of a Utah pond contaminated with toxic sludge and rusting cars, while Bill Shrout visited the EPA's Gulf Breeze, Florida, Laboratory, and Frank Lodge traveled to Ann Arbor, Michigan, to photograph low-polluting electric cars (pages 14, 58, and 33).[9]

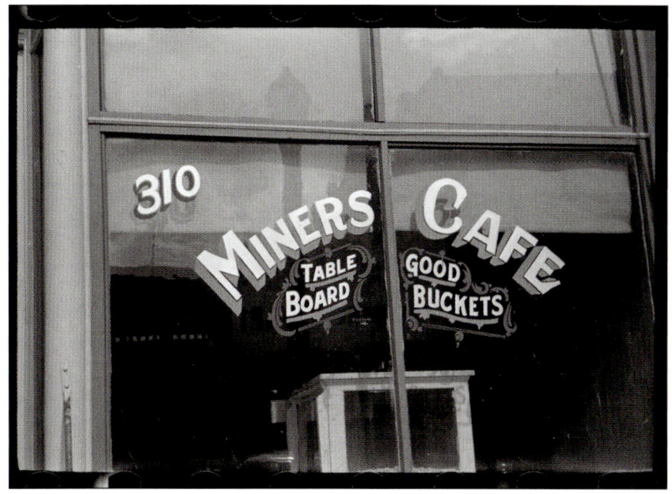

"Miner's Café, Butte, Montana, 1939"
Arthur Rothstein
Courtesy of the Library of Congress Farm Security Administration/Office of War Information Collection
(LC-USF33-003114-M2)

In keeping with the agency's mission, many assignments dealt with different forms of pollution. Marc St. Gil's photographs of Louisiana focused on chemical dumping in Lake Charles. Michael Phillip Manheim photographed the effects of noise pollution in a neighborhood near Boston's Logan Airport, while John Messina covered pesticide inspections at a chemical company in Greenville, Mississippi (pages 15, 86, and 36). But even those assignments tried to find Hampshire's "human connection." Manheim's photographs, for example, revealed the health consequences of jet noise, as well as its impact on homes near the airport, from cracks in their walls as a result of vibrations to loss of value.

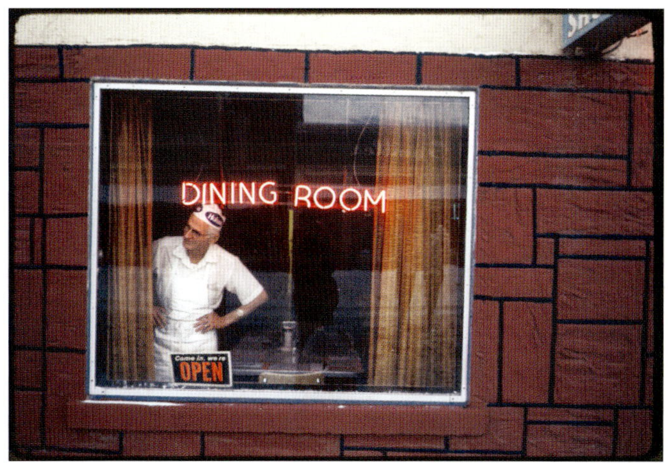

"The cook at the Texan Café watches the snow removal crew at work."
David Hiser, Rifle, Colorado, January 1973
National Archives, Records of the Environmental Protection Agency
(412-DA-10137)

More assignments came from the photographers' own interests. Lyntha and Terry Eiler's work dealt with northern Arizona, where the couple lived, and centered around "polluters on Arizona Indian Reservations." Yoichi Okamoto received an assignment to explore whether suburbia near his home in Washington, DC, was "the American dream or a gateway to frustration."[10] Flip Schulke (and later some of his photography students) returned repeatedly to New Ulm, Minnesota, where he chronicled the environmental issues in his hometown.

The most successful DOCUMERICA undertakings produced photographs that were thoughtful, artistic, or provocative. Some photographers thrived on the freedom Hampshire gave them. Leroy Woodson documented air pollution caused by steel mills in Birmingham, Alabama, but he spotlighted Birmingham's African American community and how the city's environmental problems connected race, poverty, health care, and jobs (pages 18–19, 49, and 92). Arthur Tress's photographs of Staten Island and Long Island mocked suburban lifestyles and American affluence (pages 102–103). Jack Corn's images of coal-mining communities in Appalachia documented "progressive industrial developments" in the mines, but he also photographed labor-management conflict, family life, and miners' health care issues and services (pages 38–45 and 50). Lyntha Scott Eiler's photographs of northern Arizona emphasized the clash between traditional Native American life and an energy-related economic boom (pages 110–117).

Individual DOCUMERICA photographs often brilliantly captured diverse aspects of American society in the seventies. Tom Hubbard saw Cincinnati's Fountain Square as a "public square that works for the city and its people," and many of his street photographs playfully chronicled 1970s fashion and style (pages 32 and 118–125). One Sunday morning, Marc St. Gil accompanied teenagers from Leaky, Texas, to their favorite spot to hang out and photographed them smoking pot and making out (page 30). Charles Steinhacker's shot of a young, long-haired man peering at a poster outside an X-rated movie theater in Berlin, New Hampshire, illustrated the commercialization of sexuality and how community standards were loosening (page 31). Charles O'Rear's photograph of a young hitchhiker drifting along Arizona's Route 66 with his dog, "Tripper," captured the 1970s youth culture and its challenges to traditional ideas about dress, masculinity, and work (page 52).

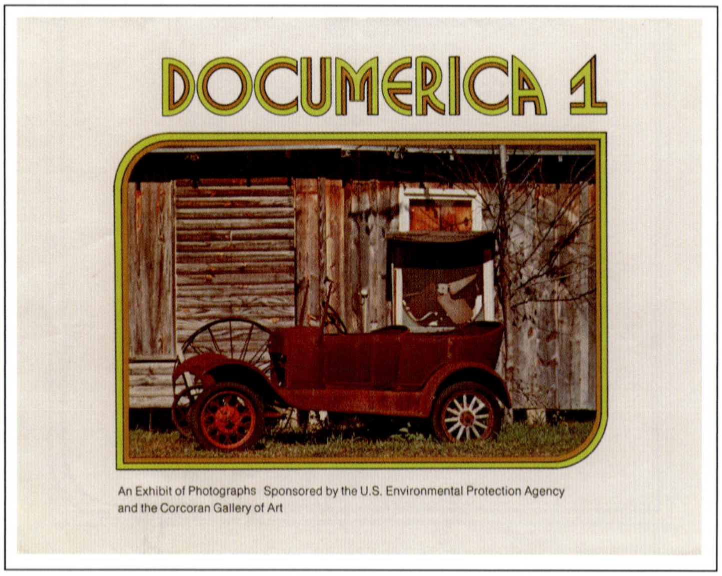

"DOCUMERICA 1," a brochure from an exhibit at
the Corcoran Museum of Art, Washington, DC, 1972
National Archives, Records of the Environmental Protection Agency

Many photographers also did an admirable job picturing the nation's growing cultural diversity, as for example Bill Gillette's series of photographs on migrant workers, Terry and Lyntha Eiler's shots of Native Americans, and Flip Schulke's study of retirees in Miami, Florida (pages 62–65 and 71). John H. White saw his extensive coverage of Chicago's African American community as a means of creating a historical record of that community, and also as a way of allowing the viewer to enter "into the lives of people you might otherwise never meet" (pages 74–81). One of Danny Lyon's assignments emphasized the contrast between "Blacks, latins, and poor whites" living in Brooklyn, New York, and "main stream America's gas station expressways shopping centers and tract homes." And a Schulke portrait pictured an interracial couple with their daughter enjoying July 4th fireworks in Sleepy Eye, Minnesota (pages 48 and 62).

Assignments and individual photographs were not always so successful. Some photographers wanted more direction from Hampshire and foundered when he refused to provide it. Others were saddled with mundane subjects and submitted repetitive images of smog, dead fish, traffic congestion, and effluence. For example, there are hundreds of aerial photographs in the project's files. While aerial photography can be a useful way to depict air pollution and suburban sprawl, many photographers fell back on this easy solution—a solution that distanced the photographer from the "human connection" Hampshire prized. Another of the project's shortcomings was that, with only a few exceptions, photographers did not take viewers inside homes. This might not seem important to a project on the environment, but such failures avoided an exploration of how individuals contributed to the environmental crisis through their lifestyles and economic decisions.

Hampshire's fascination with the Great Depression and the FSA created both benefits and drawbacks for DOCUMERICA. FSA photographers concentrated on images of 1930s rural poverty, sympathetically portraying its victims as waiting for solutions provided by New Deal agencies. At their best, they offered powerful challenges to inequality and created an unmatched visual record of American society at a moment of crisis. Drawing on the FSA for inspiration, DOCUMERICA's photographers occasionally rose to that level of social commentary. Bill Strode's photograph of a worried mother and child evacuating Louisville, Kentucky, in the wake of a 1972 liquid chlorine spill showed her fear of environmental disaster. Woodson's photographs of Birmingham's "Korea Town"—a poor African American community located in the shadow of the city's steel mills—depict its residents as dependent on the source of pollution near their homes for jobs. Jane Cooper's caption for her portrait of a woman dairy farmer connected the woman's anxiety to "low milk prices and increasing property taxes that threaten her way of life" (pages 24, 92, and 57).

But the 1970s were not the 1930s. American society had become increasingly complex, fragmented, and diverse. Cities and suburbs, not small towns and farms, dominated the American landscape; markets were global, the economy consumer-driven, and the nation's population increasingly mobile. Moreover, the solutions proposed by the EPA, and depicted in many DOCUMERICA photographs, "offered regulation, but not relief." As such, they only occasionally generated sympathy or moved a viewer to political or environmental activism. And while many environmental problems might be blamed on corporate practices or governmental inaction, others originated in American affluence and consumerism. Focusing on the harmful and destructive choices most Americans (including the photographers and Hampshire) made every day cut uncomfortably close to home. With few exceptions, such as Okamoto's study of suburbia or Tress's assessment of Staten Island, DOCUMERICA photographers did not critique America's consumer culture.[11]

DOCUMERICA's images of rural and small towns were especially problematic. America had long before become a nation of cities and suburbs, but owing to the FSA's influence, many of DOCUMERICA's assignments emphasized rural locations. Unfortunately, many of the photographs from those assignments were uncritical—even romanticized. Lee Lockwood's photograph of men looking out of a pool hall window in Eastport, Maine, or Arthur Greenberg's shot of a town hall and polling place in Hickory, Illinois, are both engaging compositions, but ones that nostalgically look back to simpler times (pages 89 and 98). More importantly, few photographers in those locations made the "human connection" to the environment that Hampshire claimed was the heart of the project.[12]

Despite its shortcomings, DOCUMERICA succeeded in several ways. First, the project's ambitious cultural and social reach deserves praise. Its photographers took more than 20,000 photographs that geographically covered the entire United States and depicted in depth many of the issues facing Americans in the 1970s. The project's regional approach and its use of local photographers gave it a strong community focus—one that reflected the growing diversity of the American people.[13] Hampshire's lack of interest in "official images" that celebrated the EPA gave some photographers the freedom they wanted to explore the unexpected, even if many others found his lack of direction frustrating. As the last large-scale Federal documentary photography project of the 20th century, its images deserve attention, greater visibility, and further study.

Because DOCUMERICA was run by the EPA, many assume that its photographers photographed only dead fish, smog, and polluted rivers. During its lifetime, the project failed to capture America's imagination, but EPA's creation of DOCUMERICA's photographic file and its preservation in the National Archives will give future generations a view of the 1970s that is broad and deep. DOCUMERICA's archive moves us farther from the stereotype of the 1970s as a time when "it seemed like nothing happened," and allows us to glimpse the nascent problems, issues, and changes that would shape our future.[14]

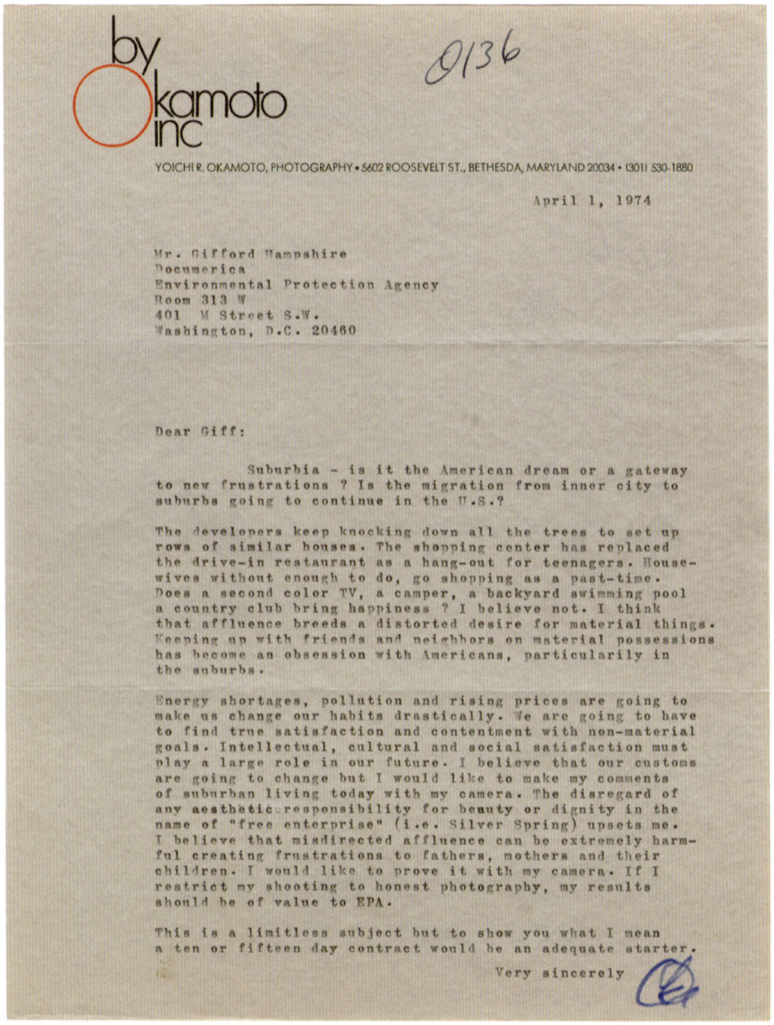

by Okamoto inc

0136

YOICHI R. OKAMOTO, PHOTOGRAPHY • 5602 ROOSEVELT ST., BETHESDA, MARYLAND 20034 • (301) 530-1880

April 1, 1974

Mr. Gifford Hampshire
Documerica
Environmental Protection Agency
Room 313 W
401 M Street S.W.
Washington, D.C. 20460

Dear Giff:

Suburbia - is it the American dream or a gateway to new frustrations ? Is the migration from inner city to suburbs going to continue in the U.S.?

The developers keep knocking down all the trees to set up rows of similar houses. The shopping center has replaced the drive-in restaurant as a hang-out for teenagers. House-wives without enough to do, go shopping as a past-time. Does a second color TV, a camper, a backyard swimming pool a country club bring happiness ? I believe not. I think that affluence breeds a distorted desire for material things. Keeping up with friends and neighbors on material possessions has become an obsession with Americans, particularily in the suburbs.

Energy shortages, pollution and rising prices are going to make us change our habits drastically. We are going to have to find true satisfaction and contentment with non-material goals. Intellectual, cultural and social satisfaction must play a large role in our future. I believe that our customs are going to change but I would like to make my comments of suburban living today with my camera. The disregard of any aesthetic responsibility for beauty or dignity in the name of "free enterprise" (i.e. Silver Spring) upsets me. I believe that misdirected affluence can be extremely harm-ful creating frustrations to fathers, mothers and their children. I would like to prove it with my camera. If I restrict my shooting to honest photography, my results should be of value to EPA.

This is a limitless subject but to show you what I mean a ten or fifteen day contract would be an adequate starter.

Very sincerely

Letter from Yoichi Okamoto to Gifford Hampshire proposing a DOCUMERICA assignment on suburbia, April 1, 1974
National Archives, Records of the Environmental Protection Agency

Opposite: "Montgomery Mall outside of Bethesda. The large shopping center has replaced the drive-in restaurant as the center of suburbia."
Yoichi Okamoto, Bethesda, Maryland, May 1974
National Archives, Records of the Environmental Protection Agency
(412-DM-0136-4-2)

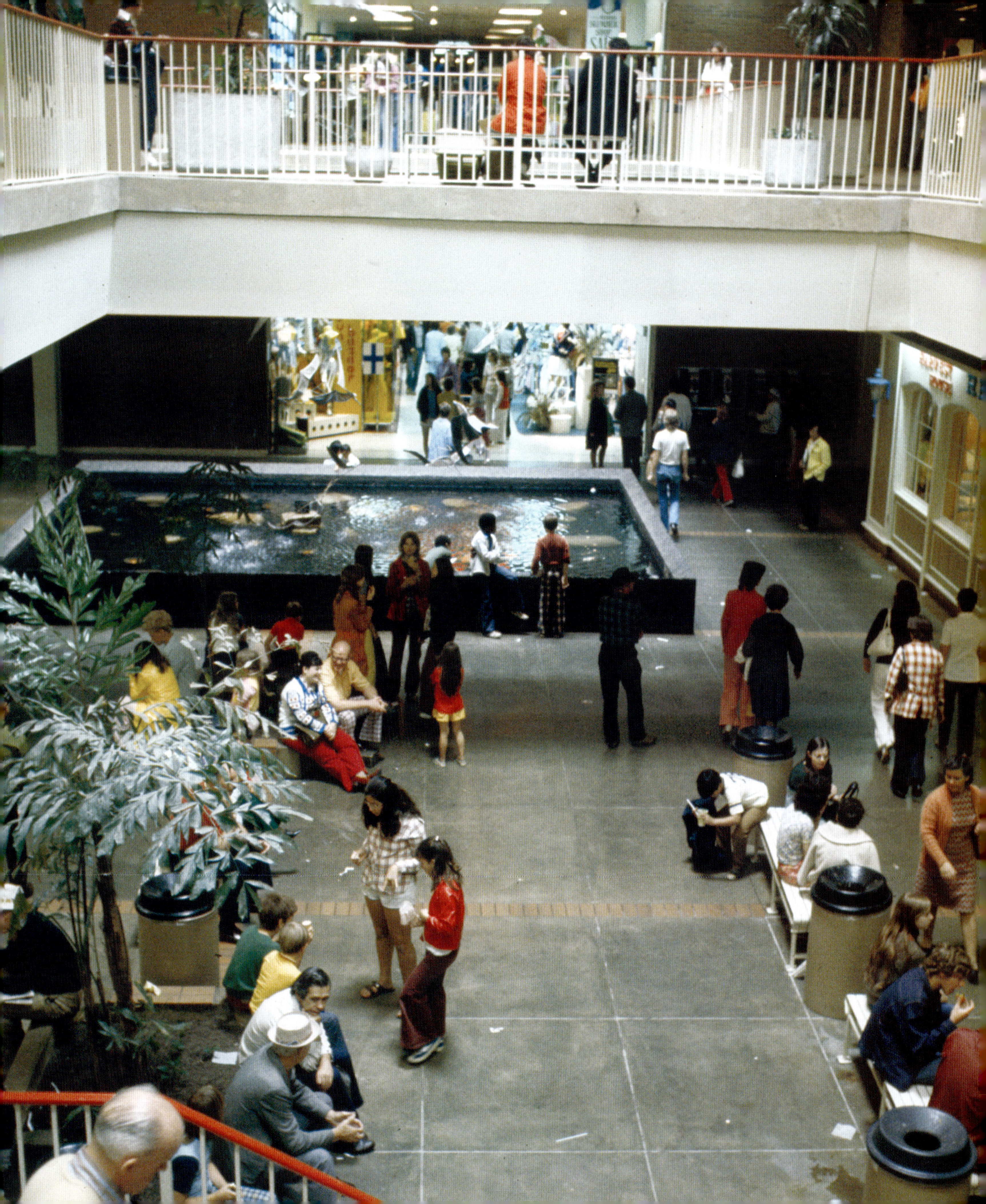

Why Are These Photographs in the National Archives?

The photographs in this book document the actions of a Federal agency—the Environmental Protection Agency. By law, when the EPA's records—and those of other Federal agencies—are no longer needed for agency business, those that have permanent value come to the National Archives and Records Administration (NARA). Only about 2–3 percent of all Federal records become part of the Archives' holdings.

When DOCUMERICA ended, Gifford Hampshire hoped his project's color slides, photographic prints, and other records would be deposited in the Library of Congress, to sit alongside the 1930s Farm Security Administration photographs he so admired. However, in 1977, the Library declined to accept DOCUMERICA's files. Worried that sending them to the National Archives would make them inaccessible, Hampshire arranged for the photographs to be deposited at the Center for Creative Photography in Tucson, Arizona. When staff members at the Archives heard the records had been sent to Tucson, they aggressively pursued those "alienated" Federal records, and, in 1981, the EPA agreed to transfer them to the Archives.

The project's records are now held in the still pictures stacks at the National Archives at College Park, Maryland. In addition to almost 22,000 color slides, black-and-white negatives, color transparencies, and photographic prints, there are 20 boxes of textual material that allow historians to study the project, its photographs, and its photographers. Those records contain correspondence with photographers, assignment files, contracts, and Hampshire's official files. They include examples of newspaper and magazine articles that used DOCUMERICA photographs and photographers' caption sheets.

The largest series of DOCUMERICA photographs—almost 16,000—have been digitized and are available through NARA's Online Public Access at *http://www.archives.gov/research/search/* Smaller collections are available on Flickr.

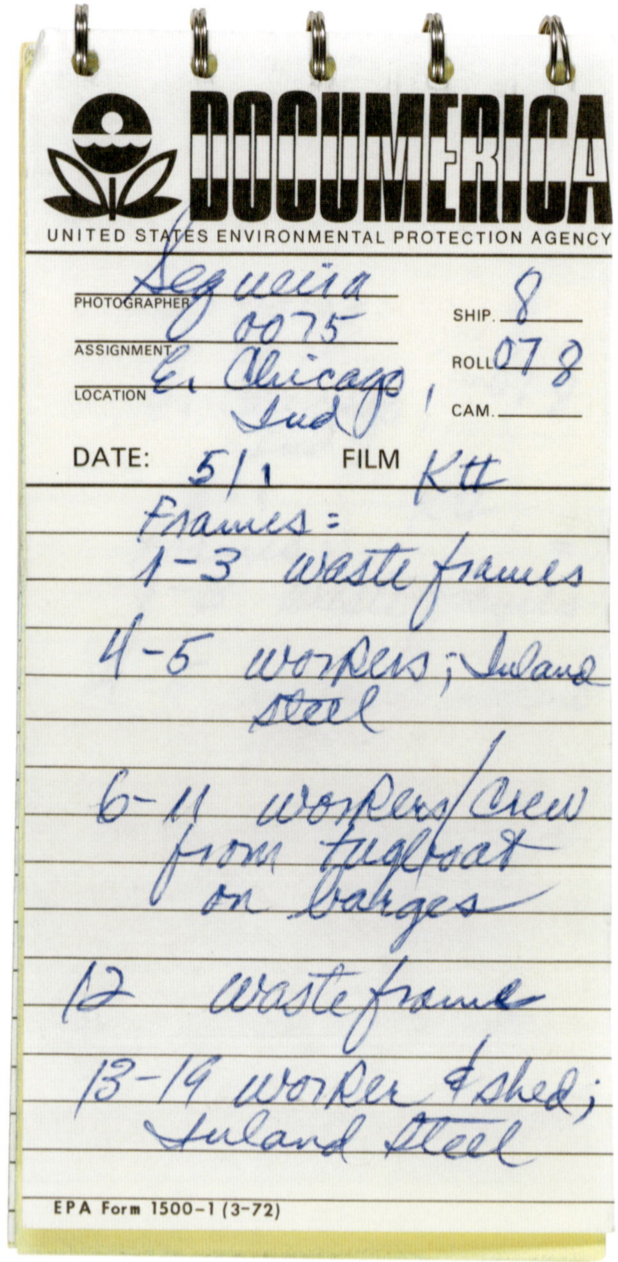

Paul Sequeira's DOCUMERICA photographer's notebook for his assignment in Chicago, Illinois, and Gary, Indiana, May 1973
National Archives, Records of the Environmental Protection Agency

Opposite: "Dockhand aboard barge."
Paul Sequeira, Chicago, Illinois, May 1973
National Archives, Records of the Environmental Protection Agency
(412-DA-4607)

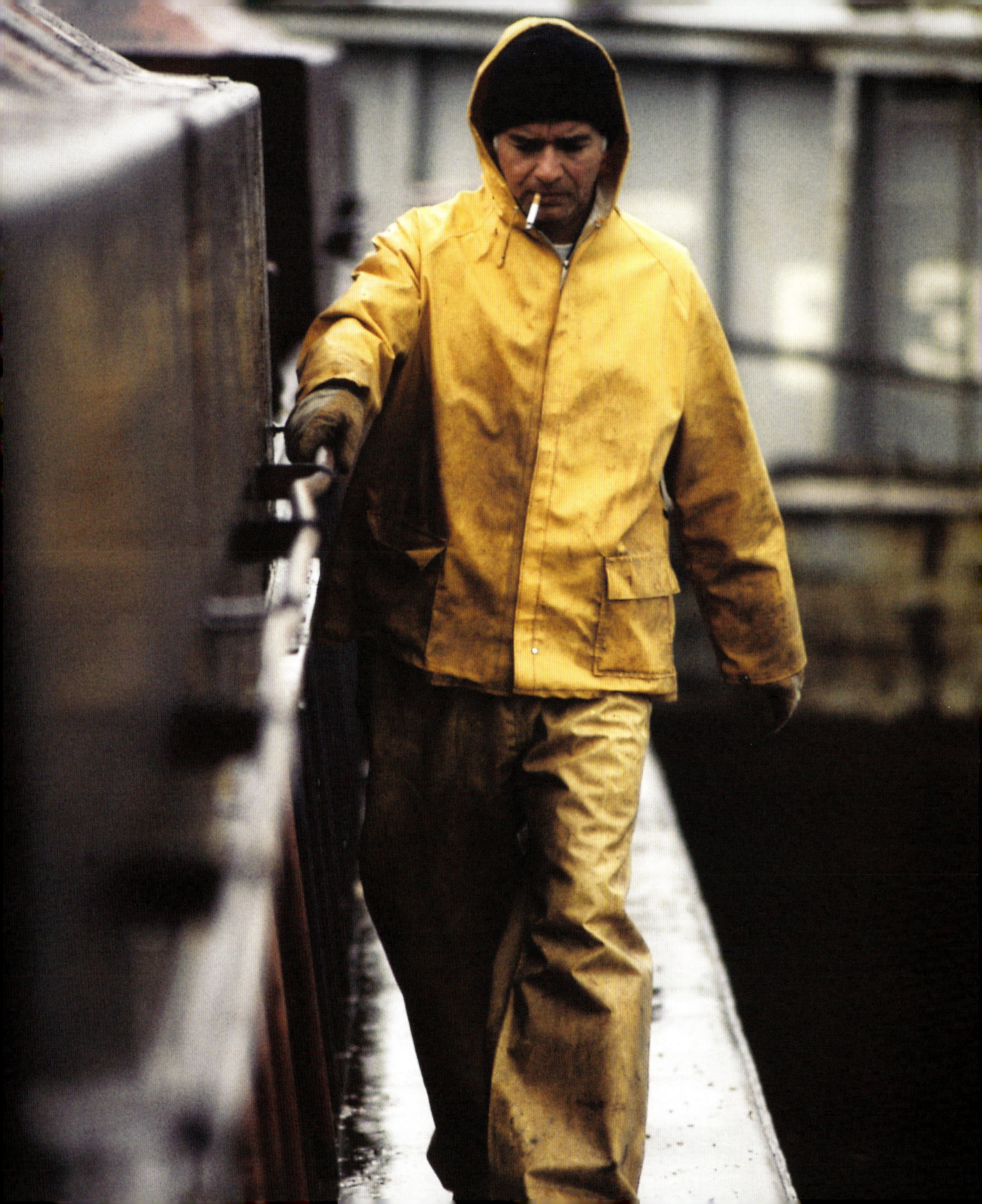

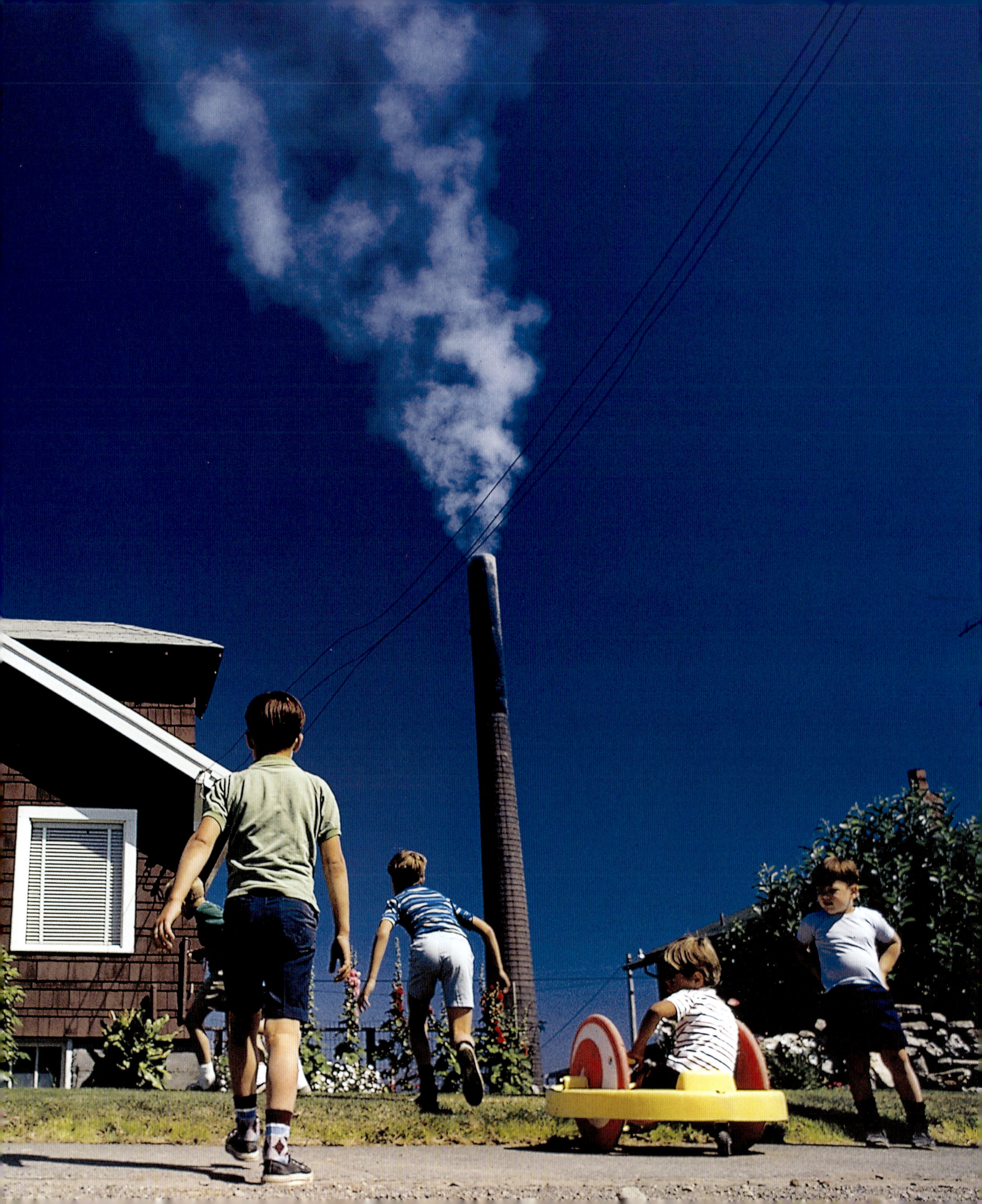

Ball of Confusion

"Ball of confusion,
That's what the world is today . . . "
"Ball of Confusion," The Temptations, 1970

The 1970s were years marked by uncertainties. The United States was powerful and wealthy, but Americans were uneasy. We were stuck economically and suffered from unemployment, inflation, and a series of energy crises. As the nation steadily changed from a manufacturing to a service-based economy, plants closed, workers protested, and jobs went elsewhere. A mass popular movement questioned the environmental costs of American affluence. Reforms meant to alleviate poverty and promote equality stalled.

Those uncertainties extended to cultural issues. We argued about how to institutionalize and extend the gains of the civil rights movement. We questioned traditional attitudes toward sexuality, marriage, and the family. We debated gender equality, abortion, drug use, gay rights, and the reach of government into our lives. Some Americans pushed for greater individual freedoms; others warned about declining moral standards.

DOCUMERICA photographers noticed and recorded many of the issues that were on the minds of seventies Americans. Along with smog, polluted waters, and suburban sprawl, they photographed urban poverty, youth culture, the women's movement, and efforts to improve the environment.

Opposite: "Children play in yard of Ruston home, while Tacoma smelter stack showers area with arsenic and lead residue."

Gene Daniels, Ruston, Washington, August 1972 (detail, see page 16)

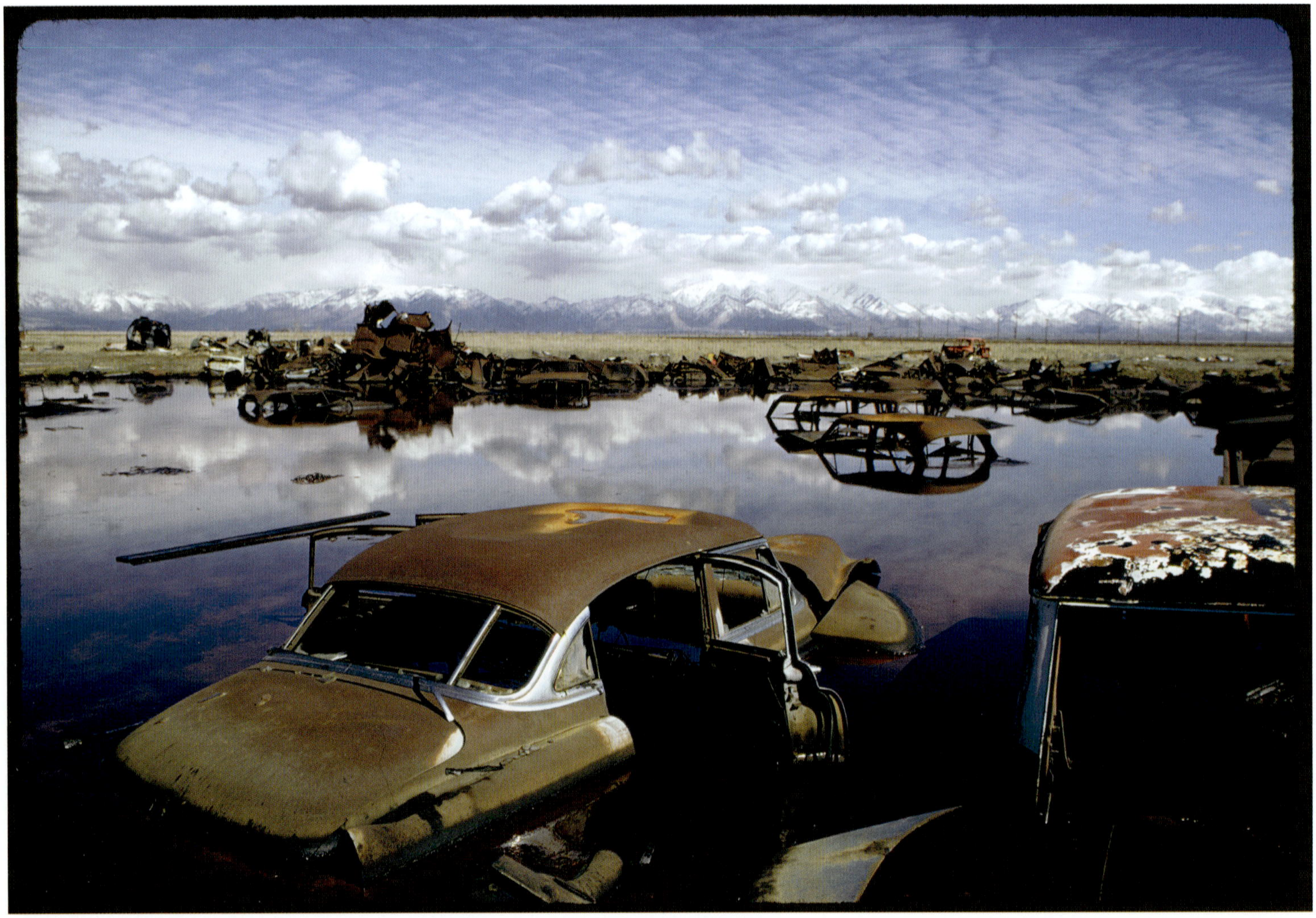

"Abandoned automobiles and other debris clutter an acid water and oil filled five acre pond. It was cleaned up under EPA supervision to prevent possible contamination of Great Salt Lake and a wildlife refuge nearby."

Bruce McAllister, near Ogden, Utah, April 1974

(412-DA-13397)

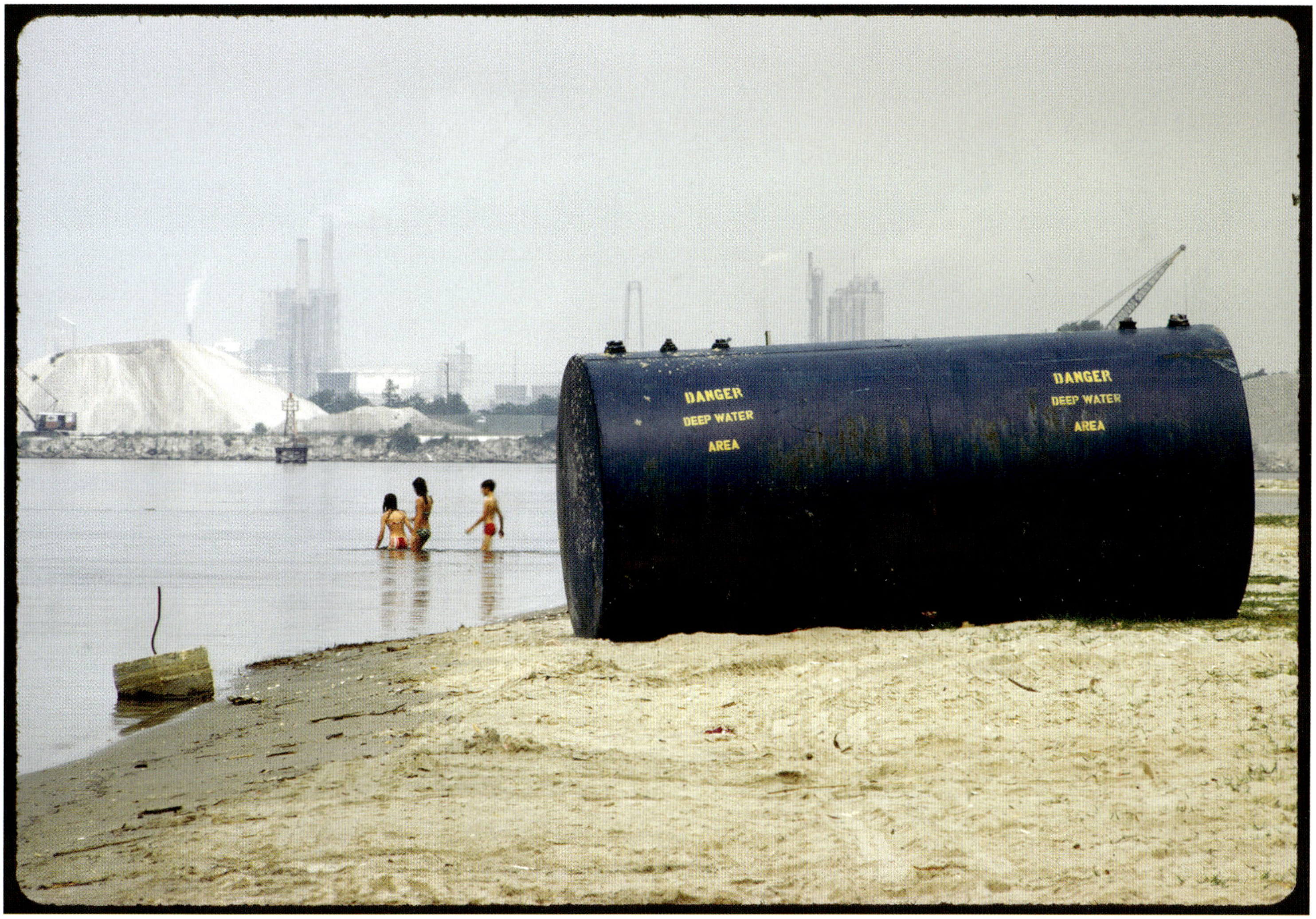

"Chemical plants on shore are considered prime source of pollution."

Marc St. Gil, Lake Charles, Louisiana, June 1972

(412-DA-3733)

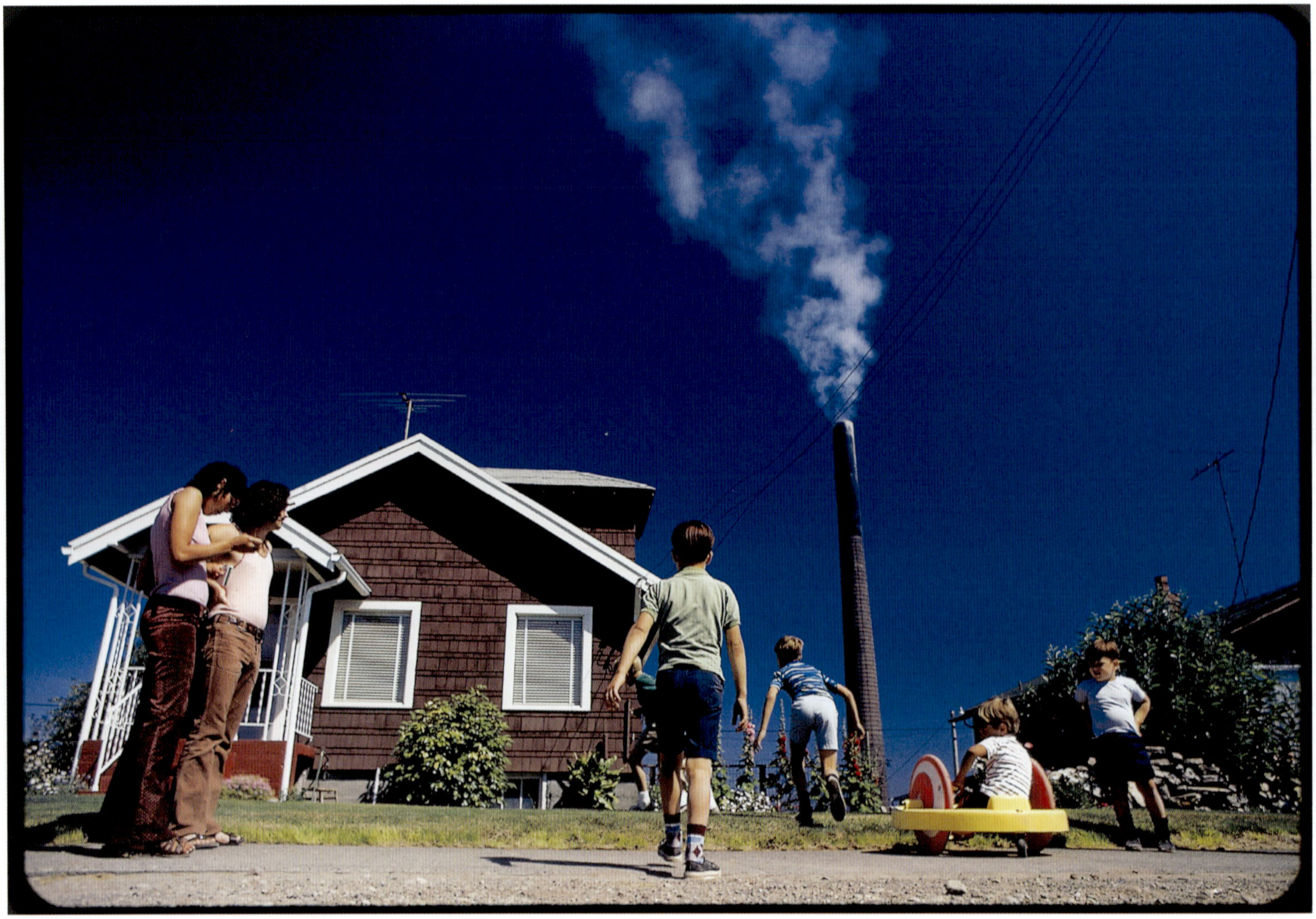

"Children play in yard of Ruston home, while Tacoma smelter stack showers area with arsenic and lead residue."

Gene Daniels, Ruston, Washington, August 1972

(412-DA-2753)

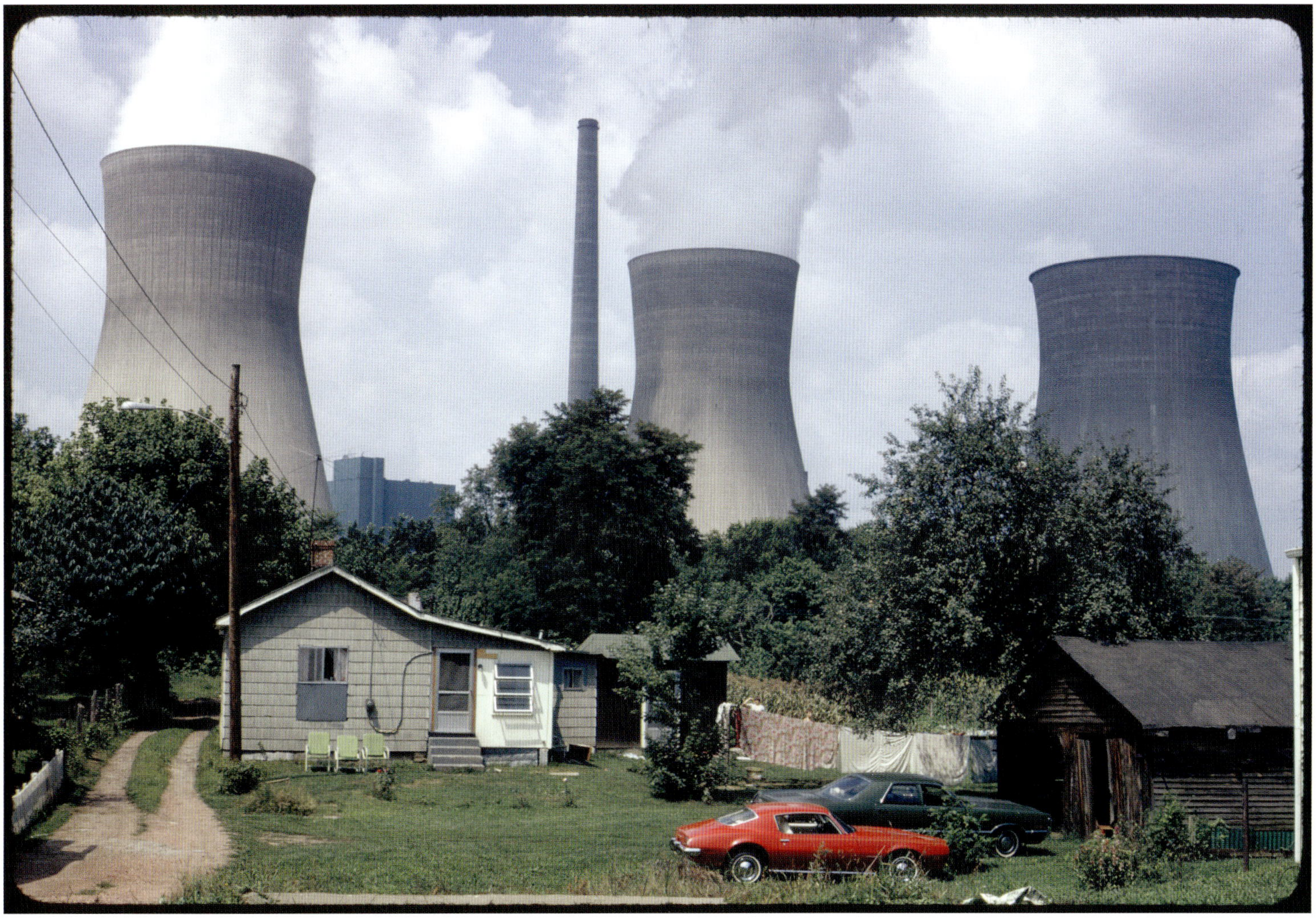

"Water cooling towers of the John Amos Power Plant loom over Poca, WV, home that is on the other side of the Kanawha River. Two of the towers emit great clouds of steam."

Harry Schaefer, Poca, West Virginia, August 1973

(412-DA-8666)

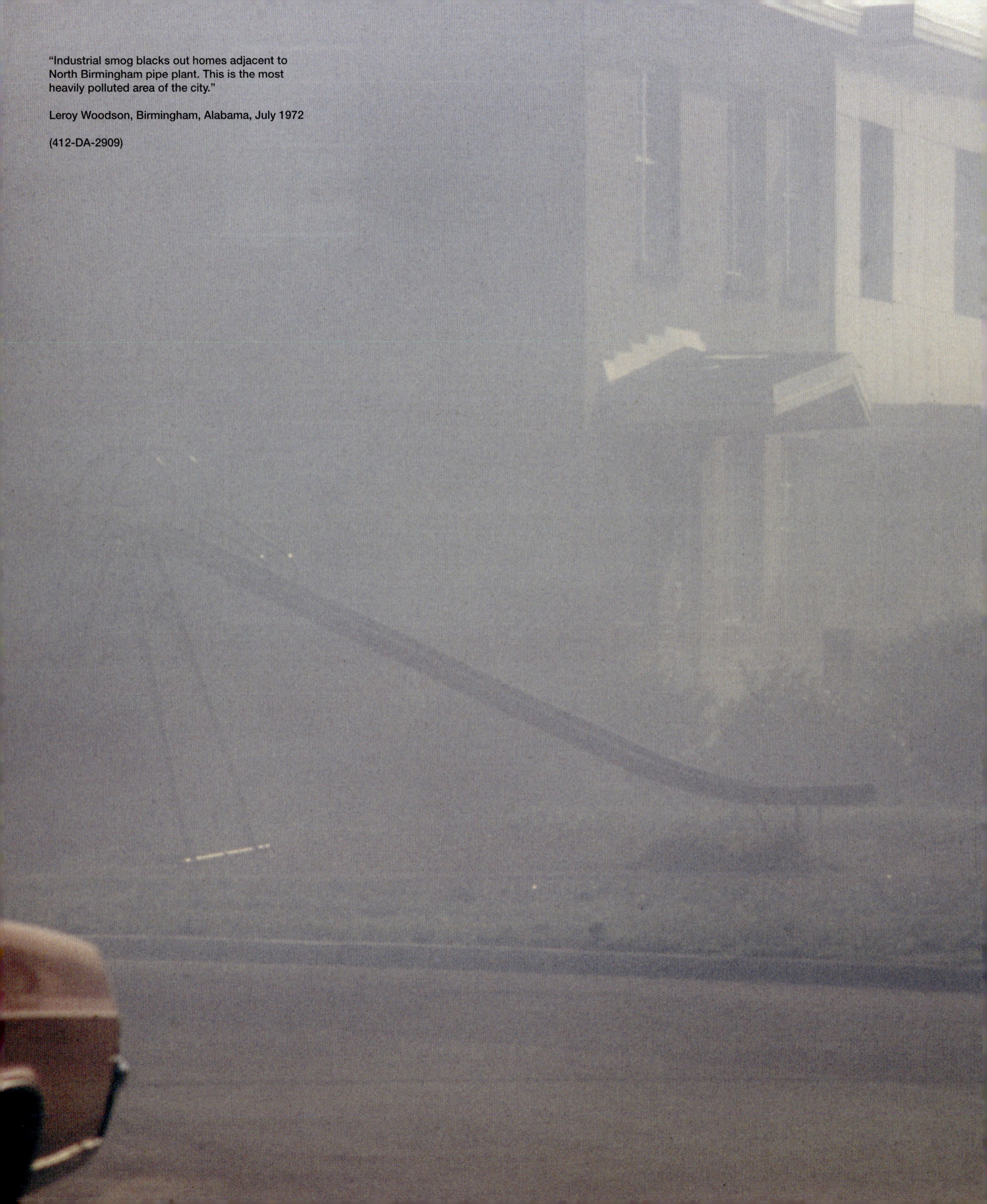

“Industrial smog blacks out homes adjacent to North Birmingham pipe plant. This is the most heavily polluted area of the city.”

Leroy Woodson, Birmingham, Alabama, July 1972

(412-DA-2909)

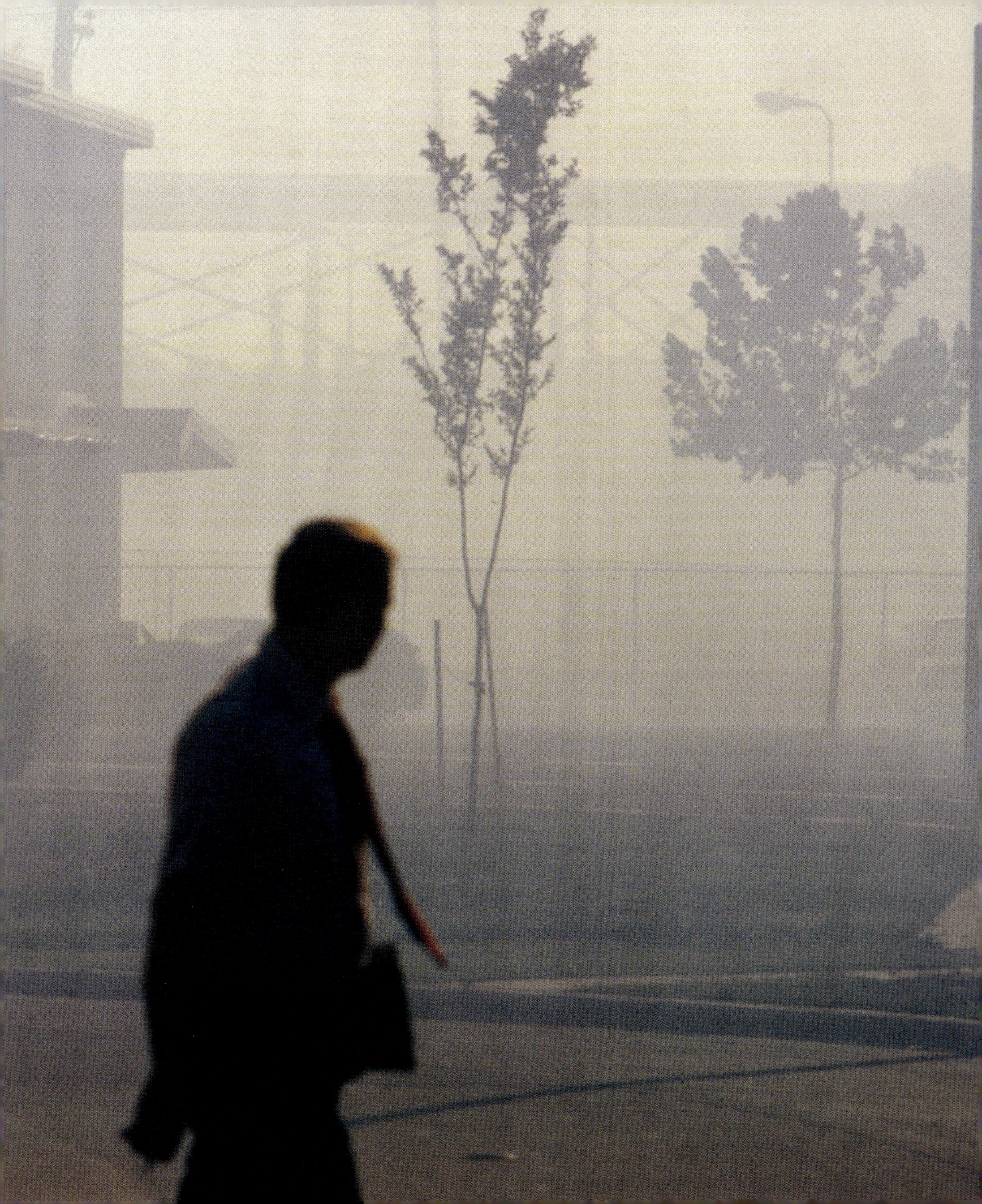

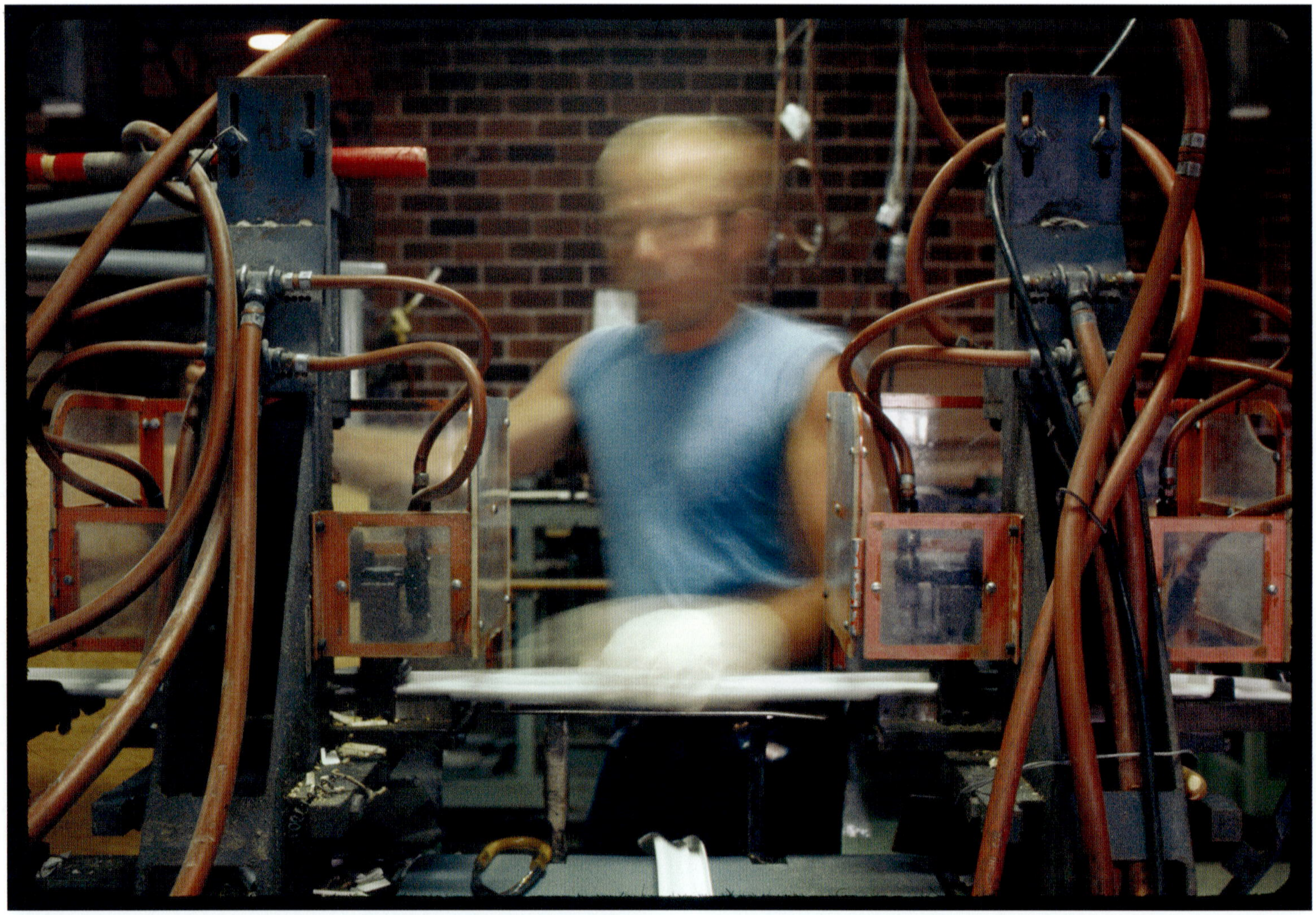

"Employee of the B.F. Goodrich Co. working on a machine that makes molding for magnetic doors seals for various brands of refrigerators. The firm employs 184 people which makes it the third largest employer in town behind the 3M Co. (Minnesota Mining and Manufacturing) with 1,377 and Kraft foods with 487. New Ulm is a country [*sic*] seat of 13,000 in a farming area of south central Minnesota founded in 1854 by a German immigrant land company."

Flip Schulke, New Ulm, Minnesota, July 1974

(412-DA-15930)

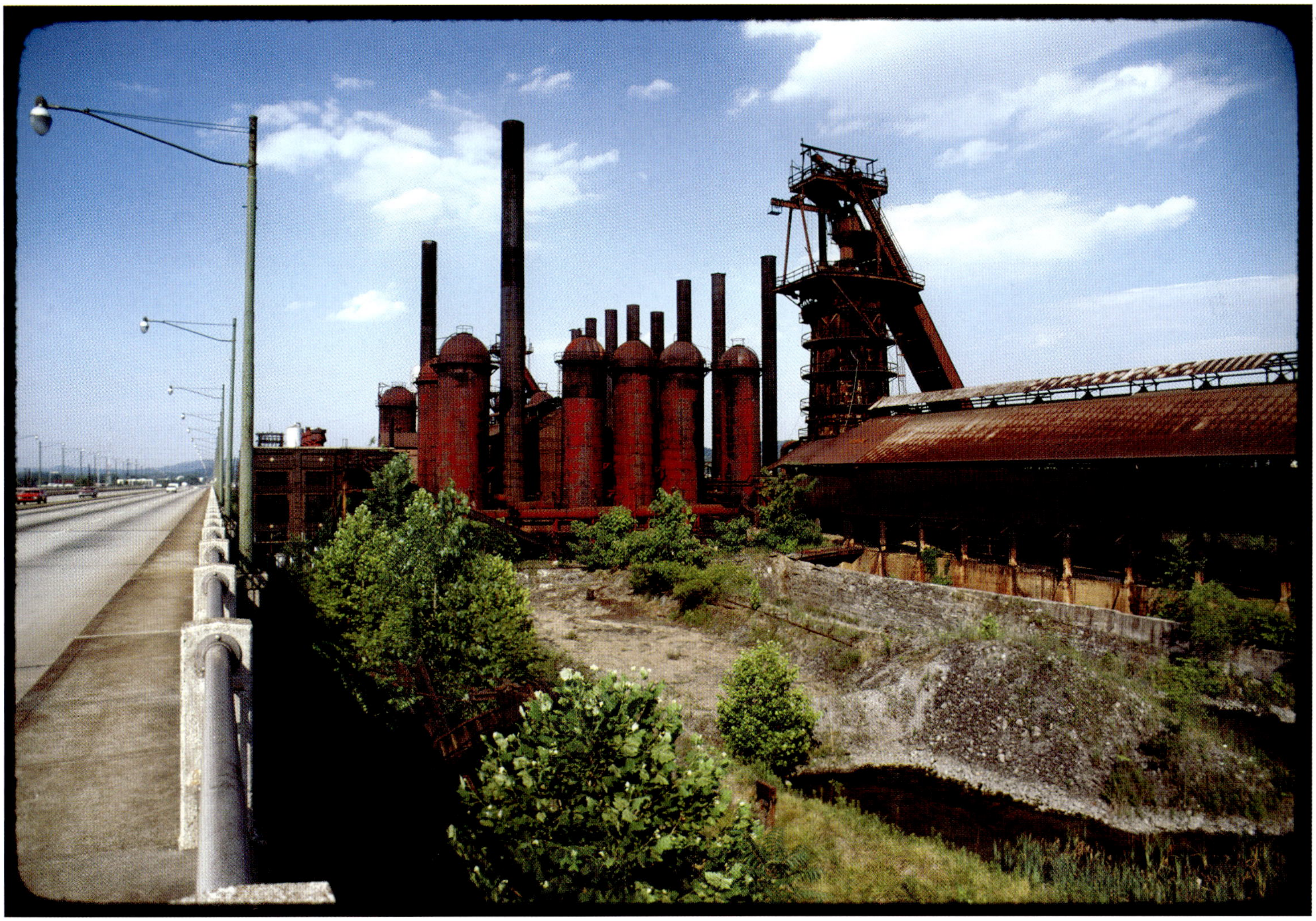

"U.S. Pipe Co., Birmingham. Closed by EPA. View from 1st Ave., No[rth] overpass. Plant now owned by city of Birmingham, may be turned into a museum."

Al Stephenson, Birmingham, Alabama, June 1977

(412-DM-0166B-20)

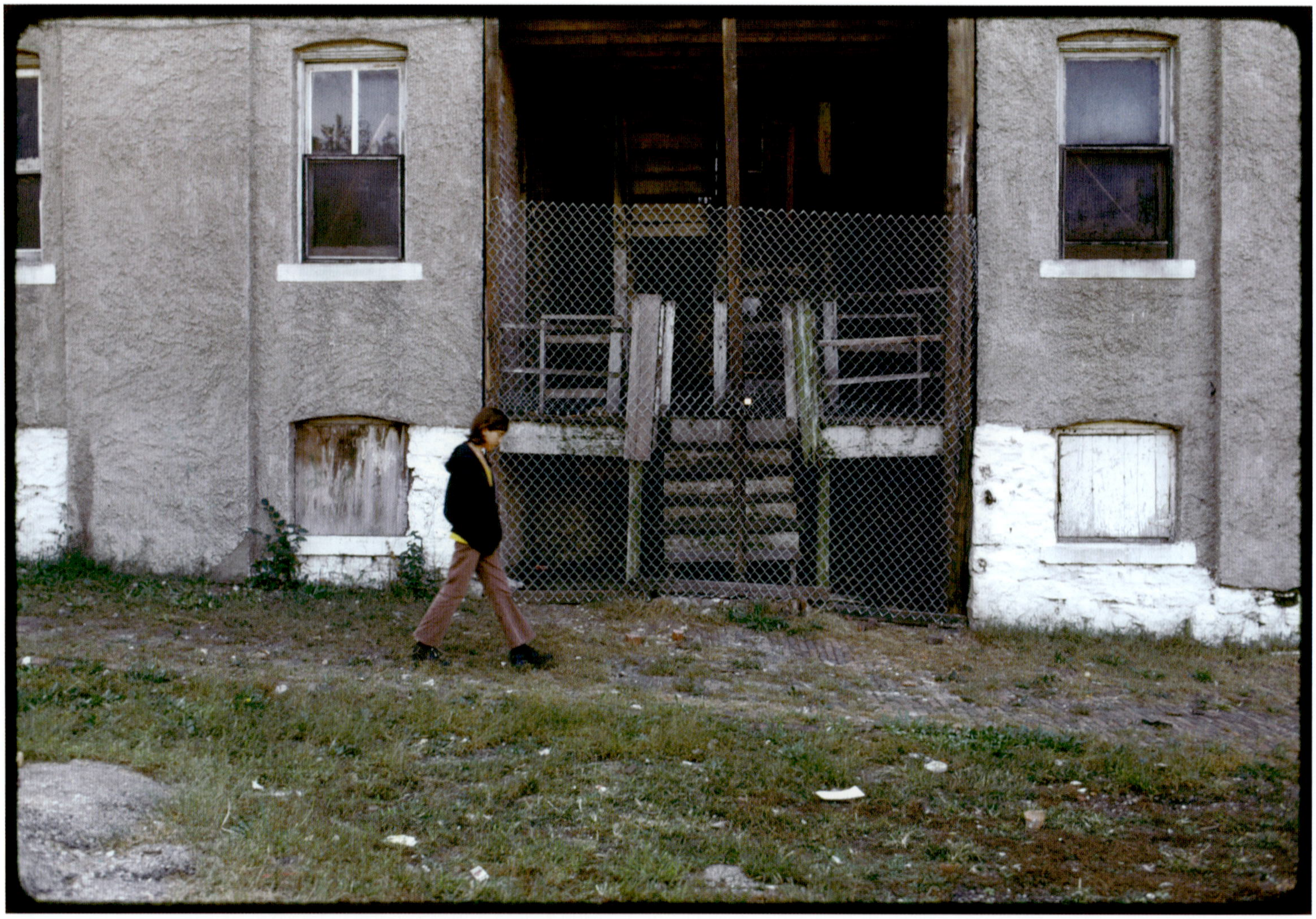

"Residents of Mulk[e]y Square, a low-income neighborhood on the city's west side, face large scale displacement by a new interstate highway (I-635). Many houses have been demolished, many others stand empty. For the people who remain, life goes on with little change. Billy Watkins, one of the children of a biscuit worker's family, passes a run-down apartment building on his way home. The tenants have gone, the building is closed."

Kenneth Paik, Kansas City, Missouri, October 1973

(412-DA-11080)

Opposite: "A black man who is jobless sits on the windowsill of a building in a high crime area on Chicago's South Side. He has nothing to do and nowhere to go. This scene contrasts with the publications which list the city as the 'black business Mecca of the world.' In early 1975 some 16% of blacks were believed to be out of work, double the rate of white unemployment. Black owned businesses in Chicago in 1970 grossed $332 million from 8,750 businesses."

John H. White, Chicago, Illinois, July 1973 (caption written in 1975)

(412-DA-13704)

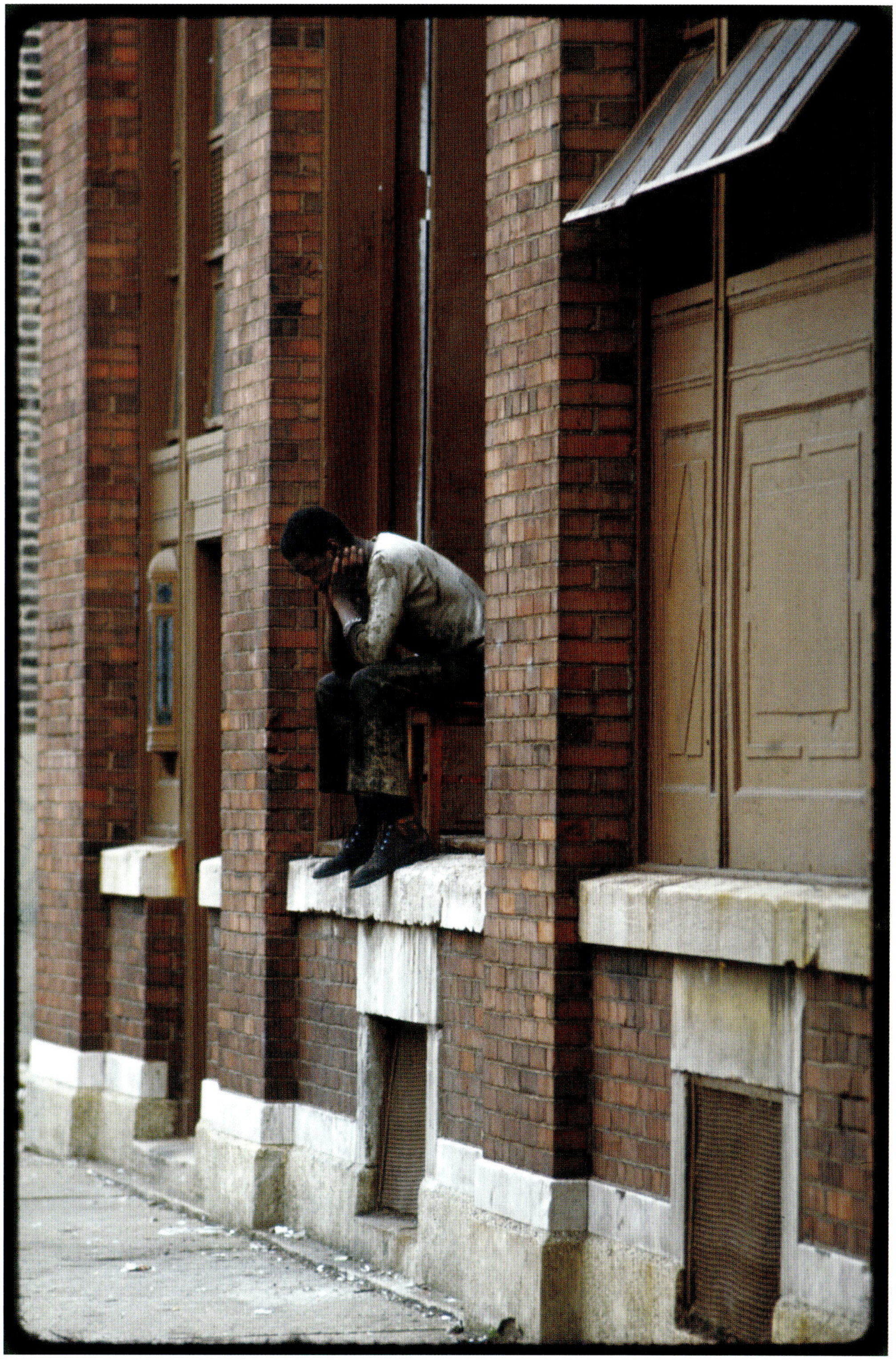

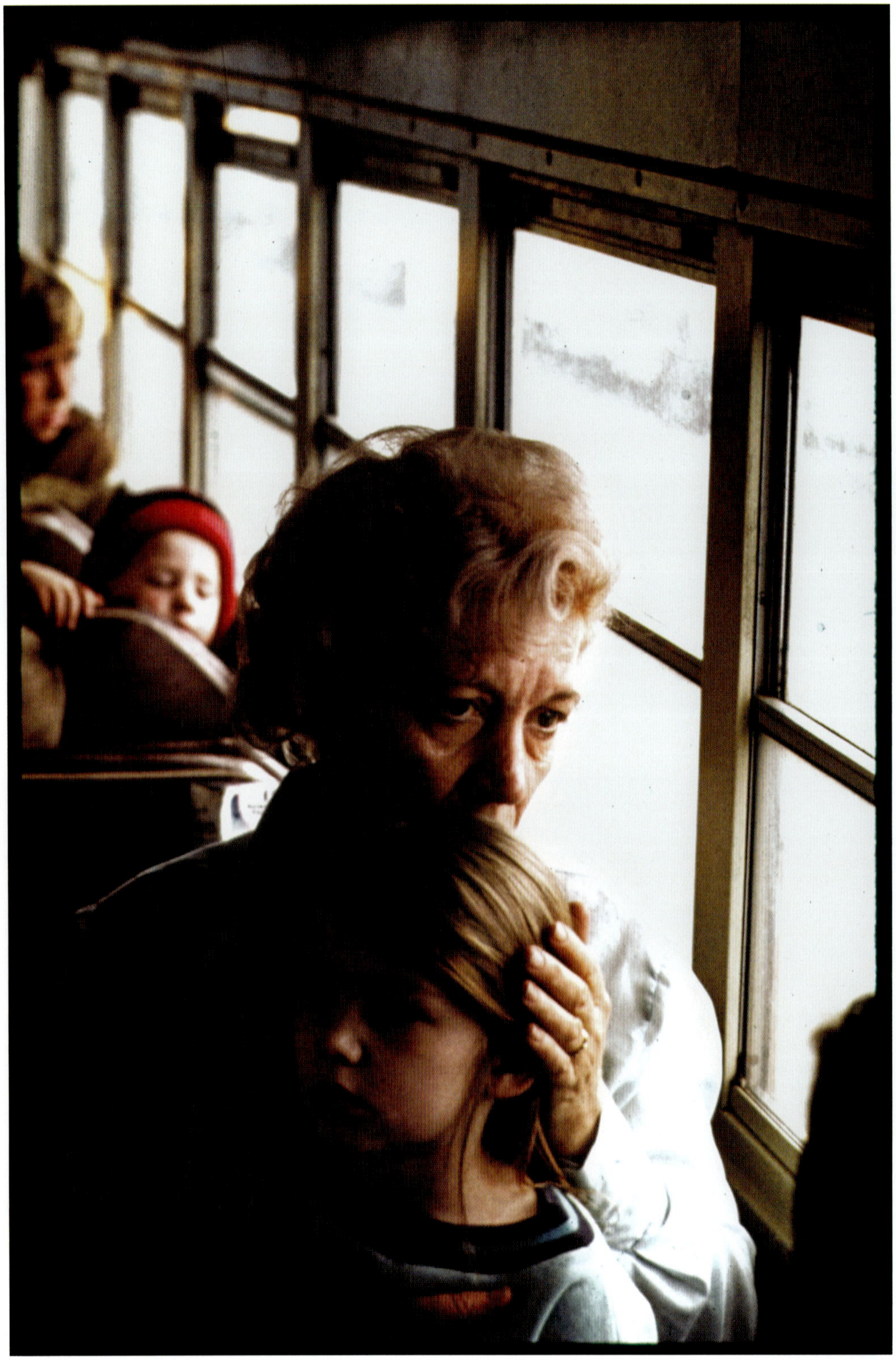

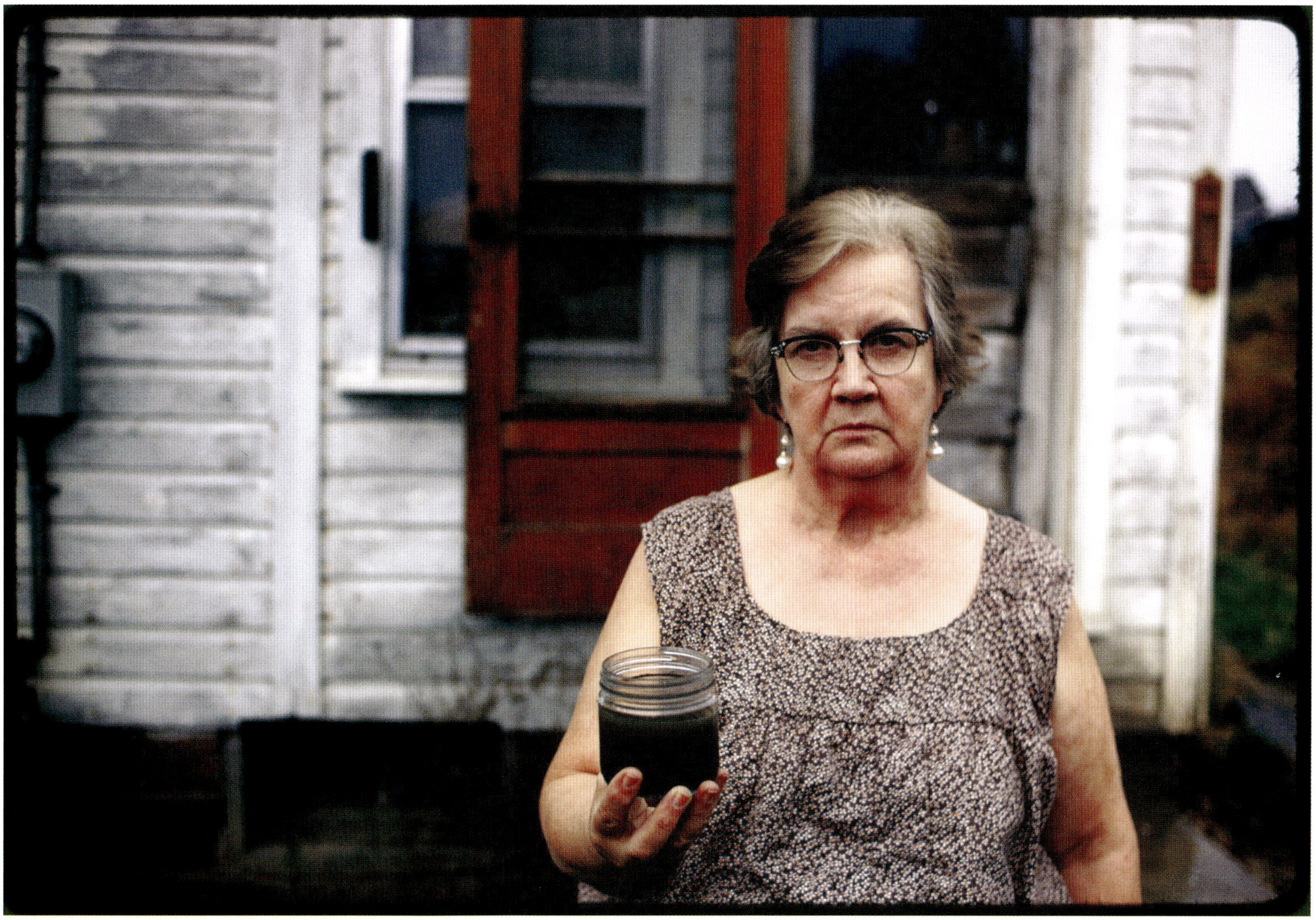

"Mary Workman holds a jar of undrinkable water that comes from her well, and she has filed a damage suit against the Hanna Coal Company. She has to transport water from a well many miles away although the coal company owns all the land around her, and many roads are closed, she refuses to sell."

Erik Calonius, near Steubenville, Ohio, October 1973

(412-DA-12346)

Opposite: "A worried mother and her child leave Louisville by bus when the city was evacuated. Evacuation centers were set up when a barge carrying liquid chlorine broke loose from its tug and threatened to spill its poisonous contents. The barge carried enough liquid chlorine to devastate much of the city of Louisville."

Bill Strode, Louisville, Kentucky, March 1972

(412-DA-1405)

“Central expressway leading south into Dallas.”

Bob Smith, Dallas, Texas, May 1972

(412-DA-5277)

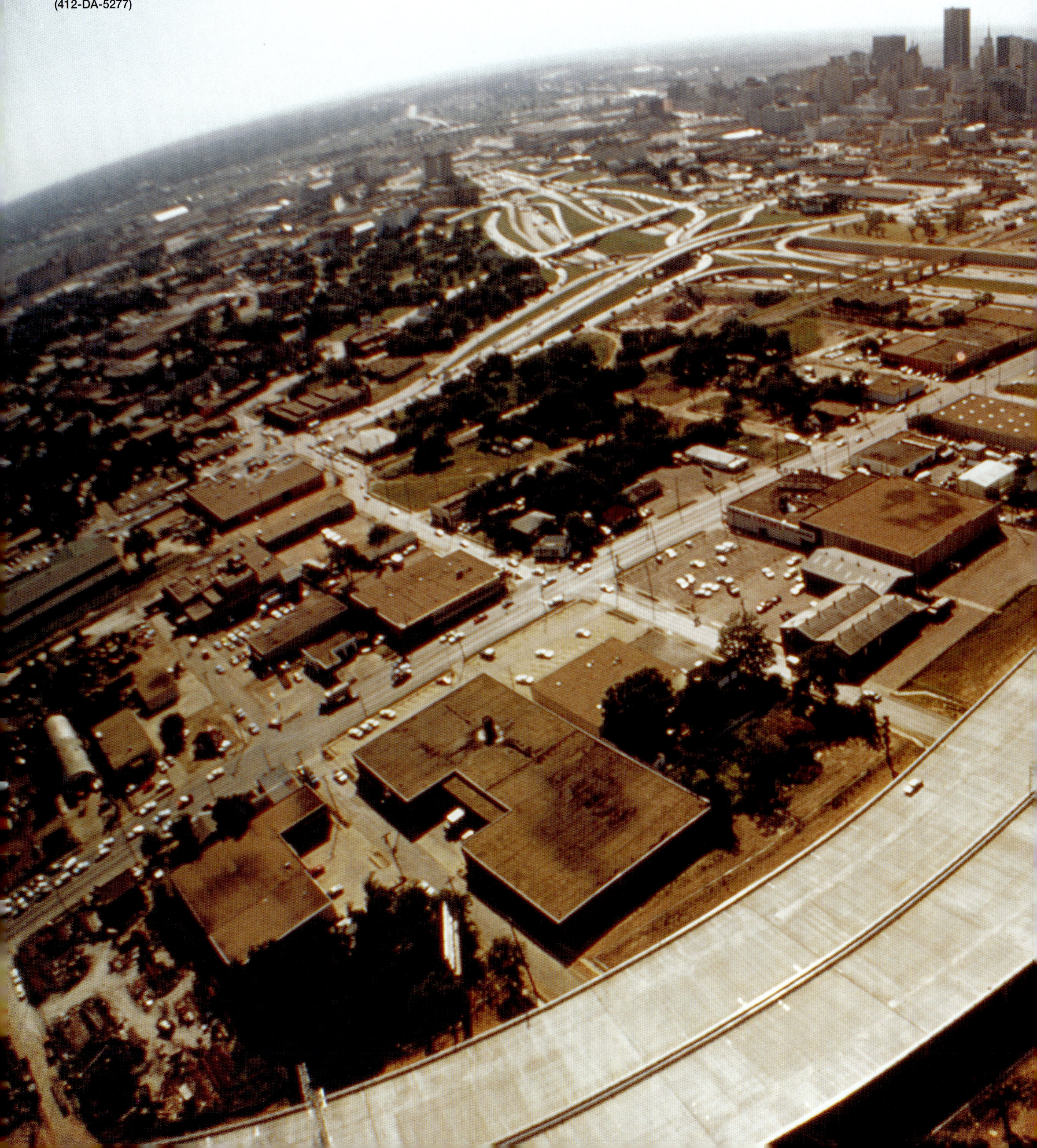

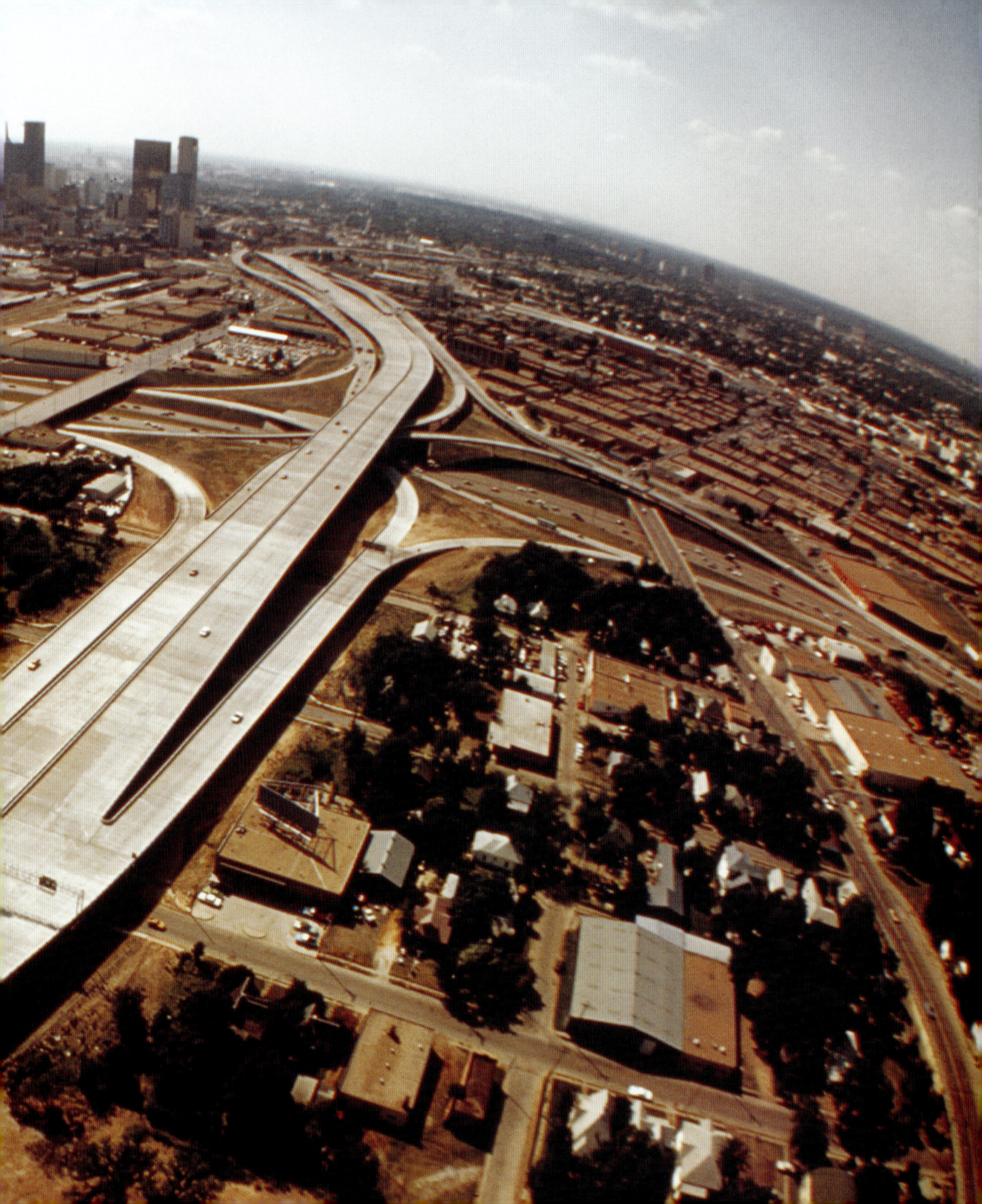

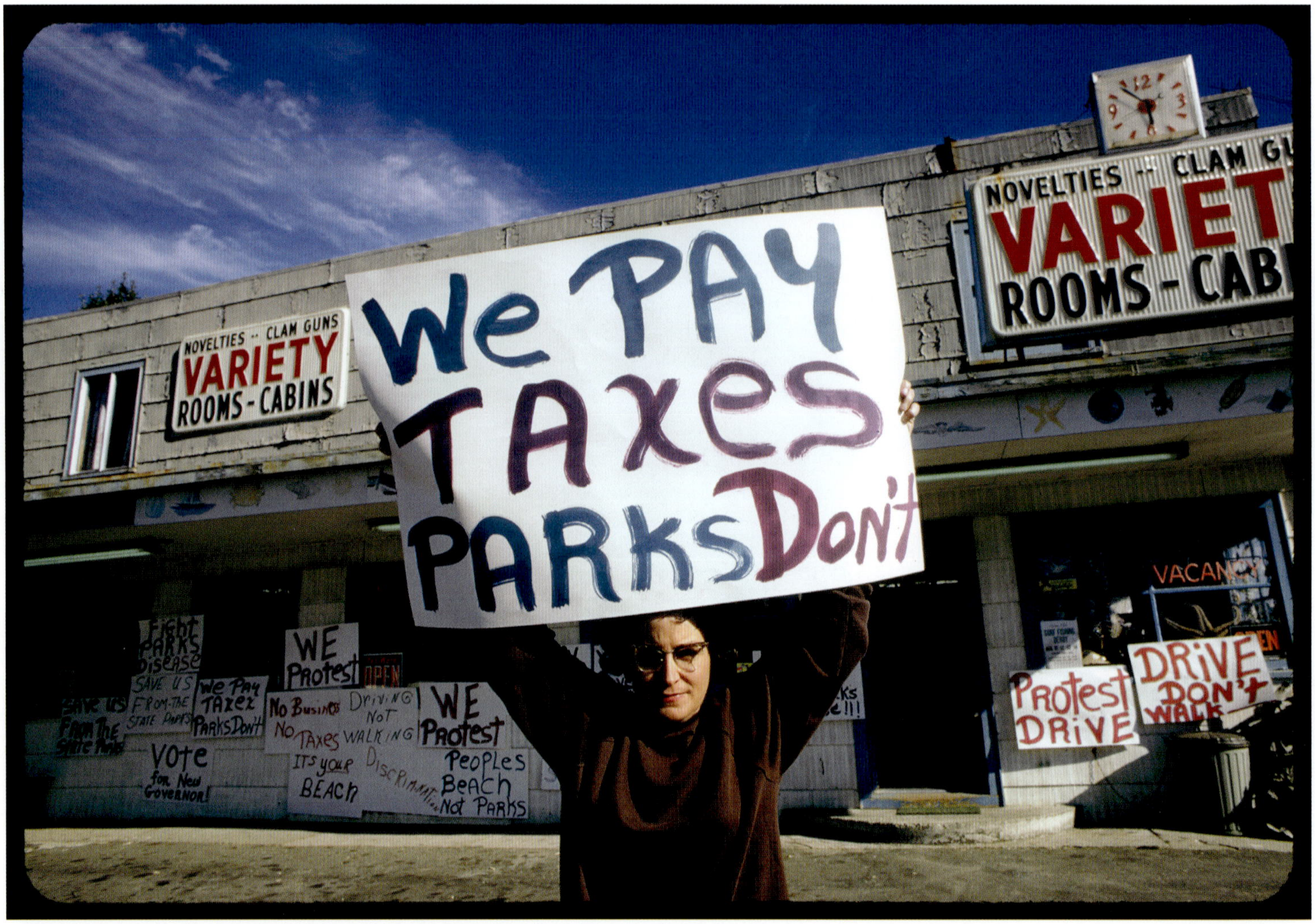

"Dorothy Thierolf, Ocean Beach businesswoman and leader of the fight to reopen nearby beach to auto traffic. To protect clam beds the state government had banned cars from a short stretch of beach during the summer months on August 12, 1972. Ms Thierolf led a demonstration in which 200 cars drove two miles through the prohibited section of the beach to protest the ban."

Gene Daniels, Ocean Beach, Washington, August 1972

(412-DA-2760)

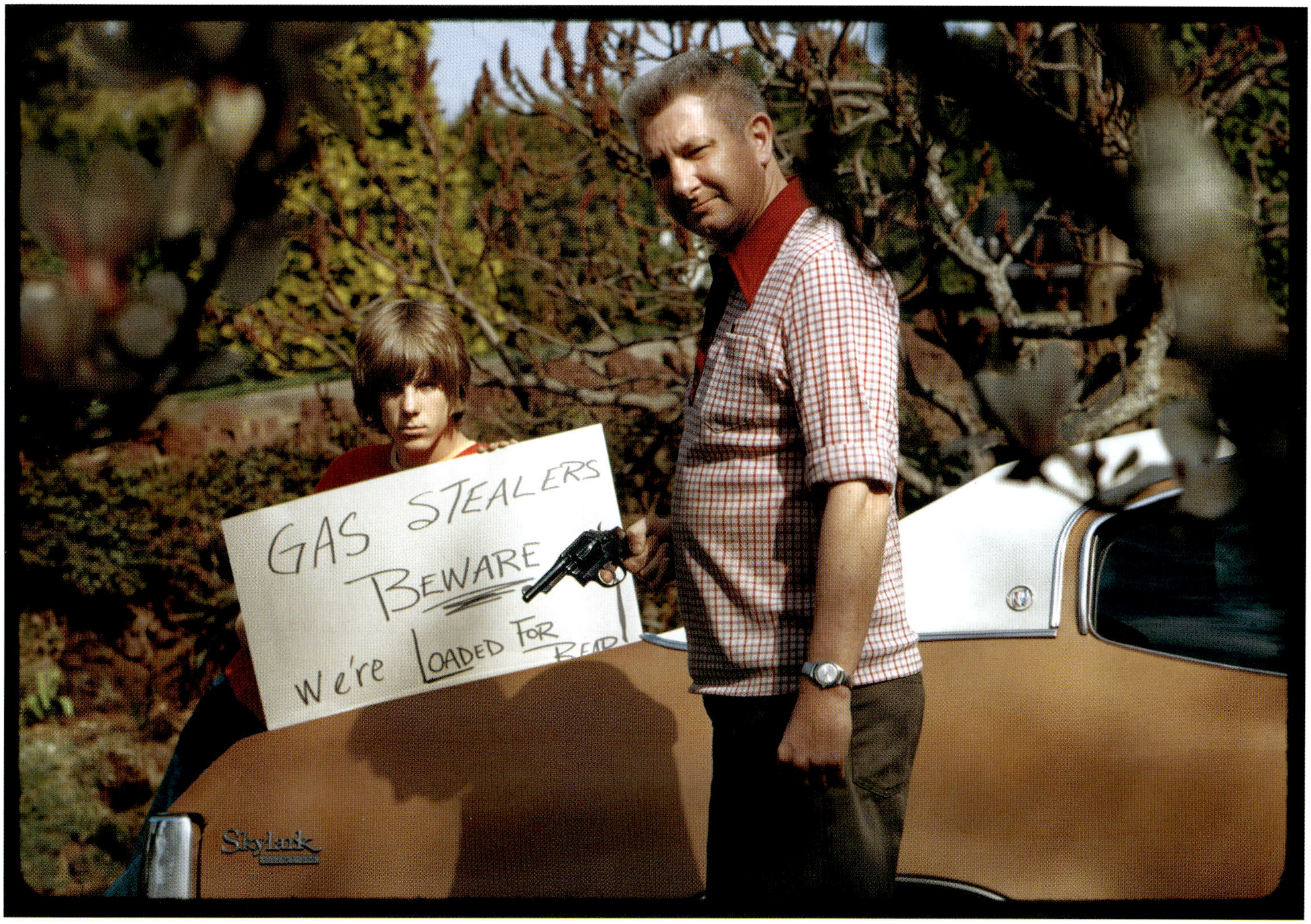

"Country's fuel shortage led to problems for motorists in finding gas as well as paying much more for it, and resulted in theft from cars left unprotected. This father and son, made a sign warning thieves of the possible consequences."

David Falconer, Portland, Oregon, April 1974

(412-DA-13063)

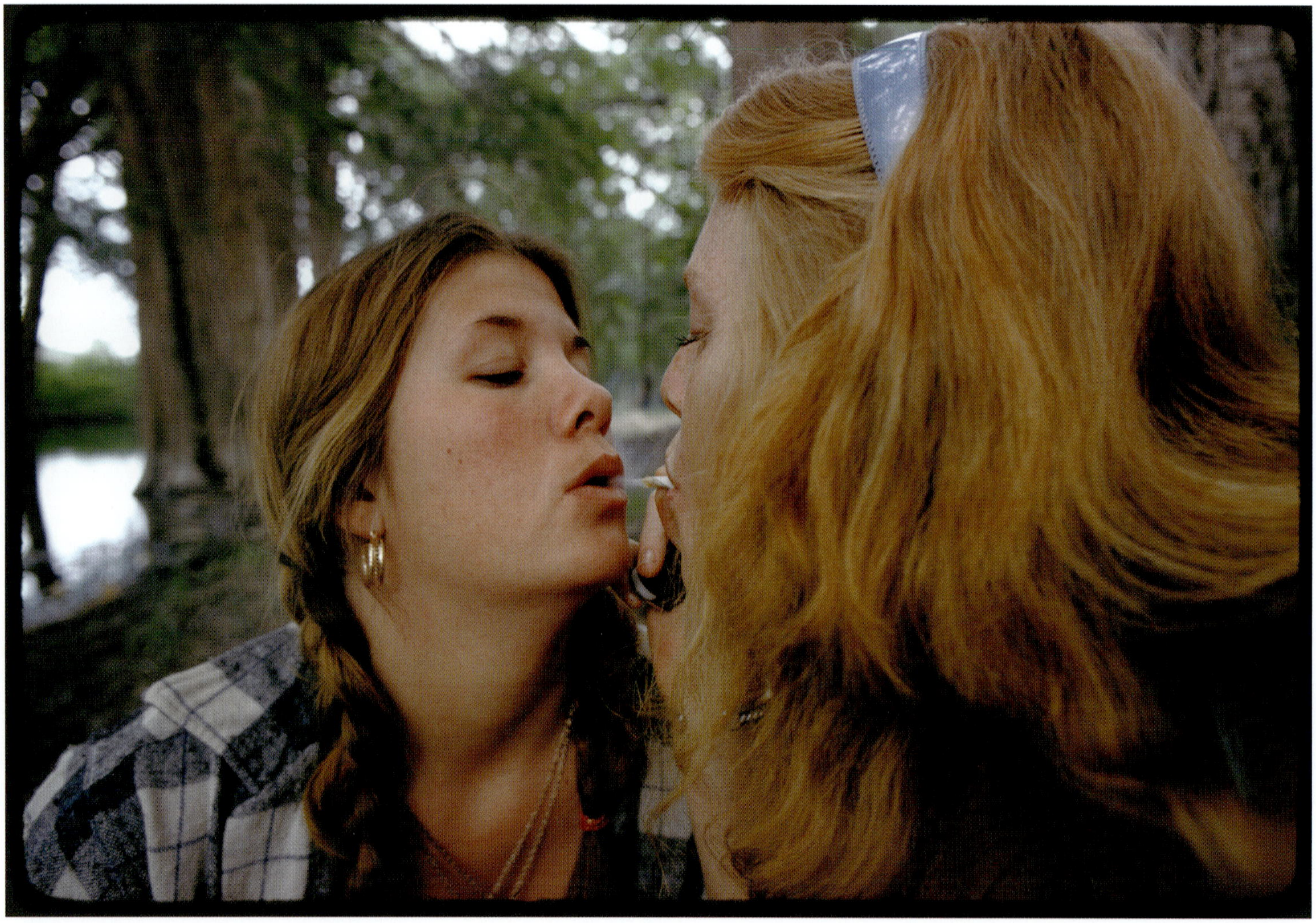

"Two girls smoking pot during an outing in Cedar Woods near Leakey, Texas. (Taken with permission.) One of nine pictures near San Antonio."

Marc St. Gil, Leakey, Texas, May 1973

(412-DA-12455)

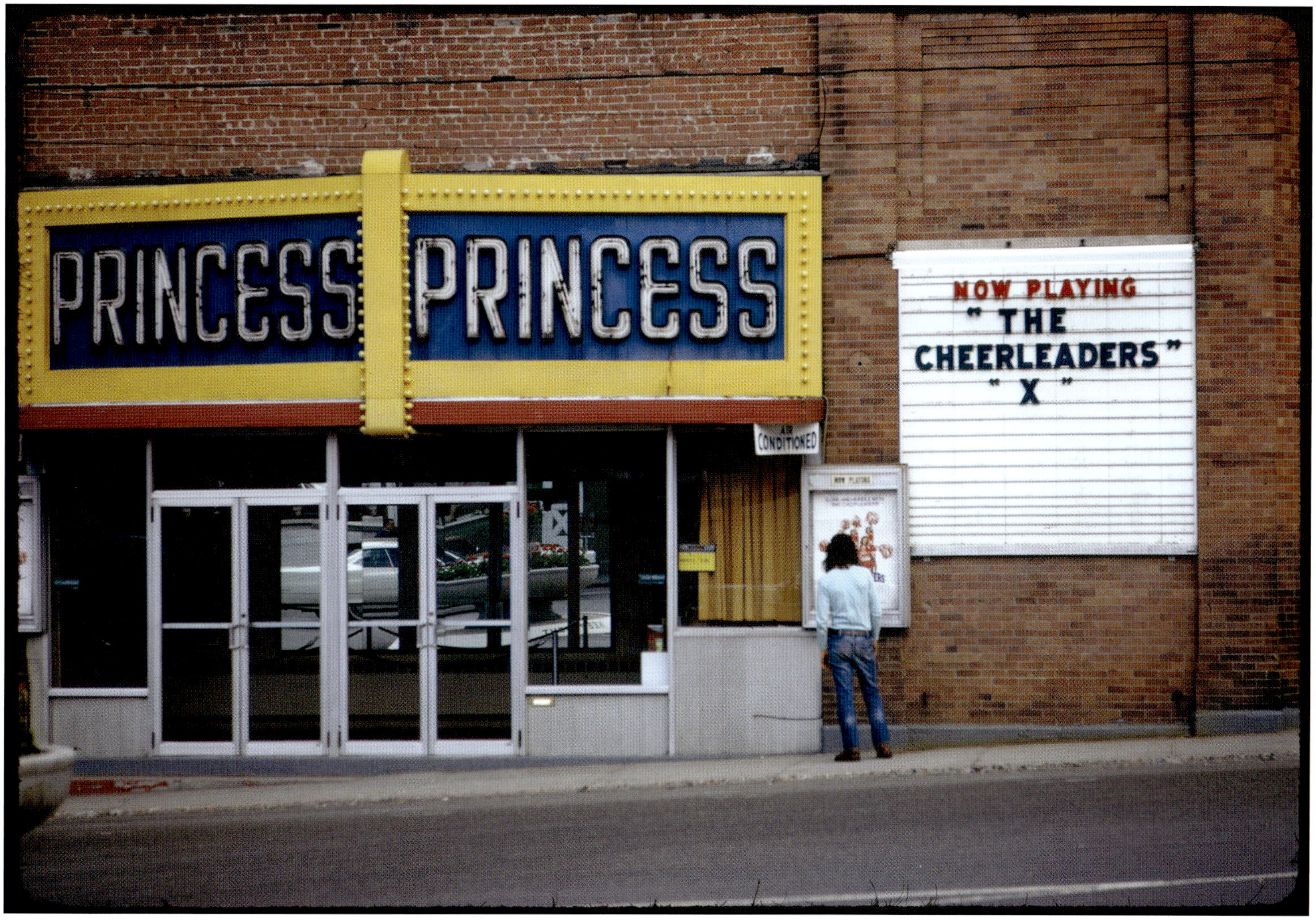

"Movie Theatre."

Charles Steinhacker, Berlin, New Hampshire, June 1973

(412-DA-8200)

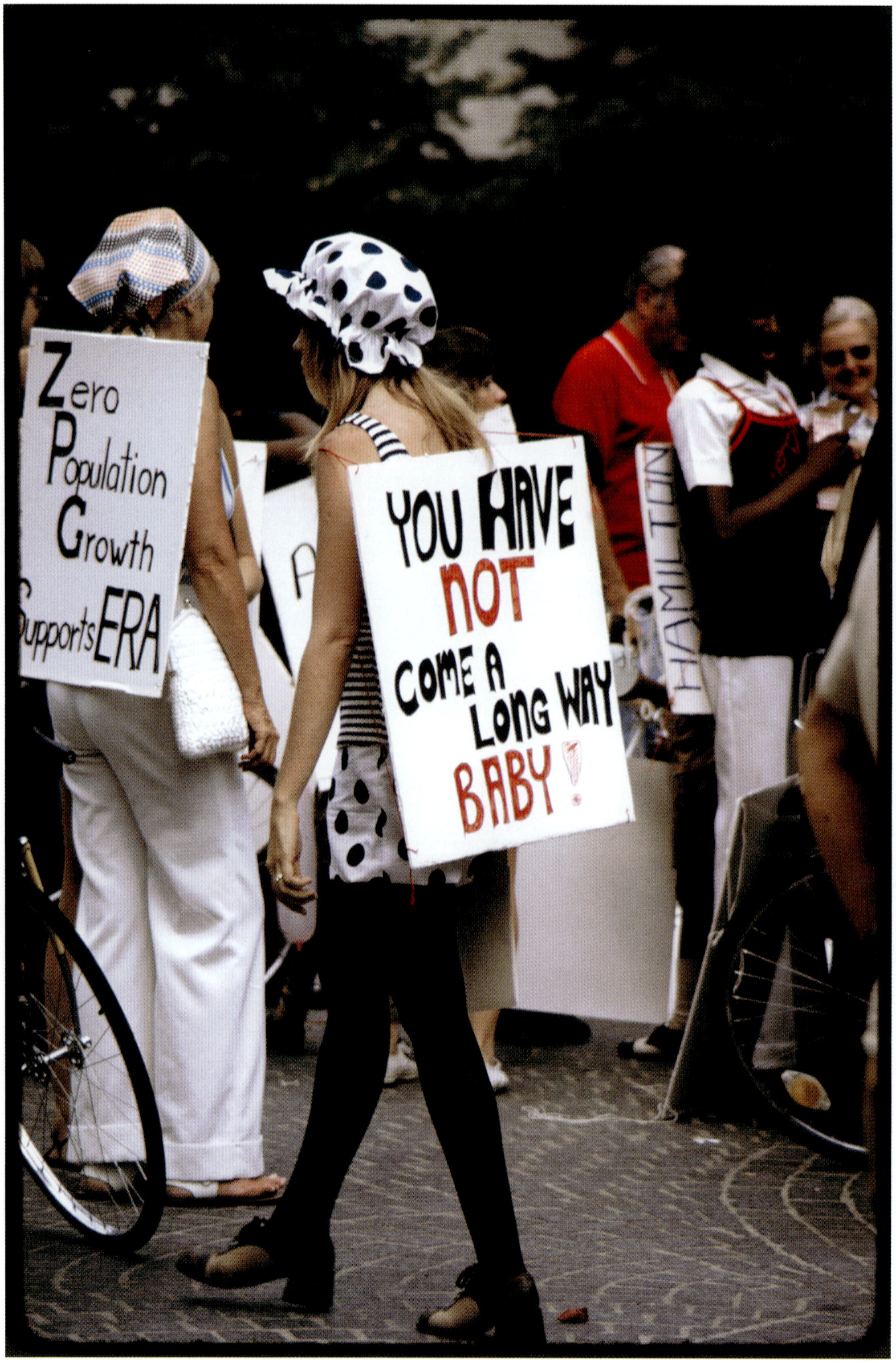
Zero
Population
Growth
Supports ERA
YOU HAVE
NOT
COME A
LONG WAY
BABY!
HAMILTON

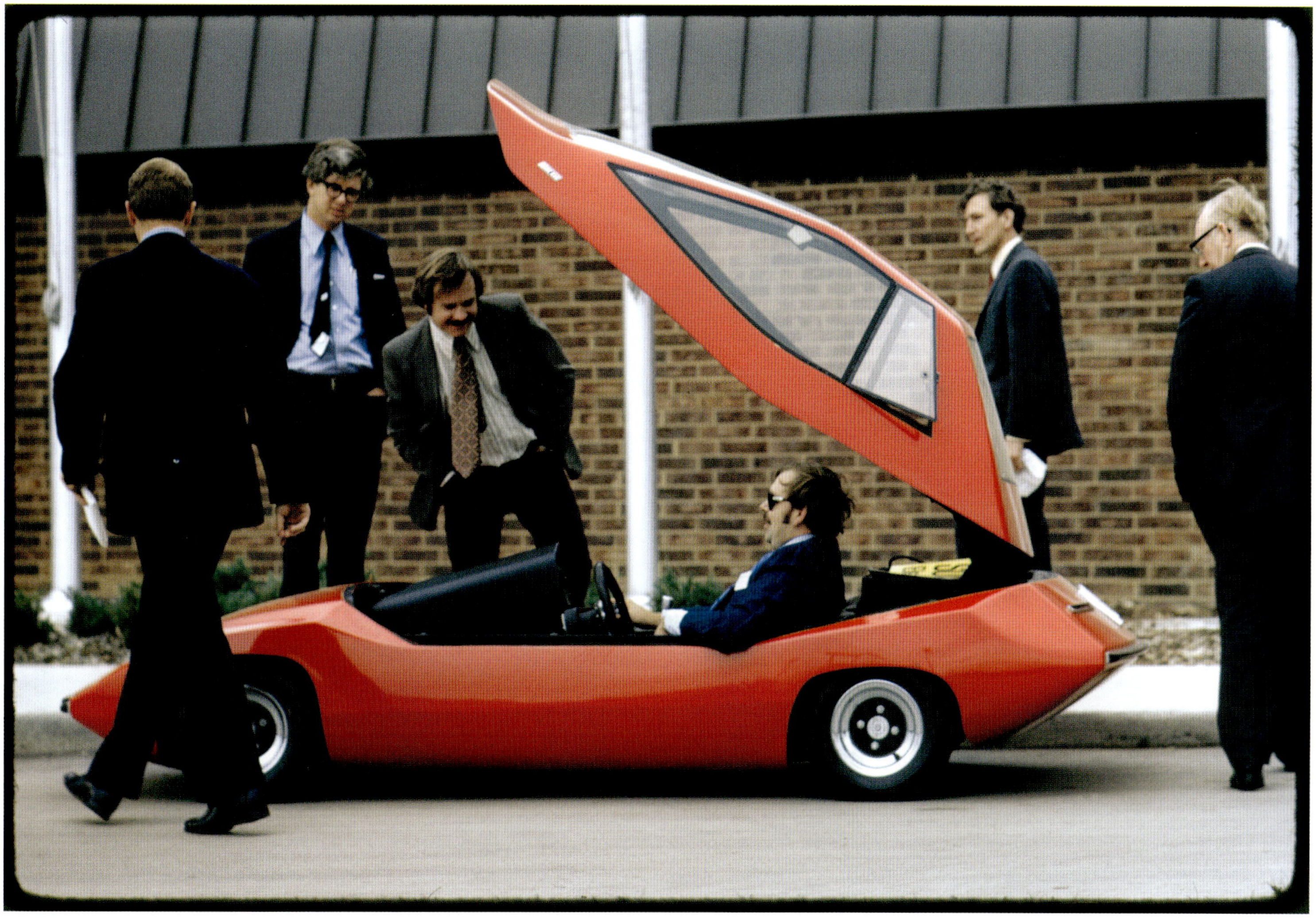

"Exhibit at the first symposium on low pollution power systems development held at the Marriott Motor Inn, Ann Arbor. Vehicles and hardware were assembled at the EPA Ann Arbor Laboratory. Part of the exhibit was held in the motel parking lot. Photo shows participants looking over the ESB 'Sundancers,' an experimental electric car."

Frank Lodge, Ann Arbor, Michigan, October 1973

(412-DA-10305)

Opposite: "Women's Suffrage Day in Fountain Square."

Tom Hubbard, Cincinnati, Ohio, August 1973

(412-DA-10848)

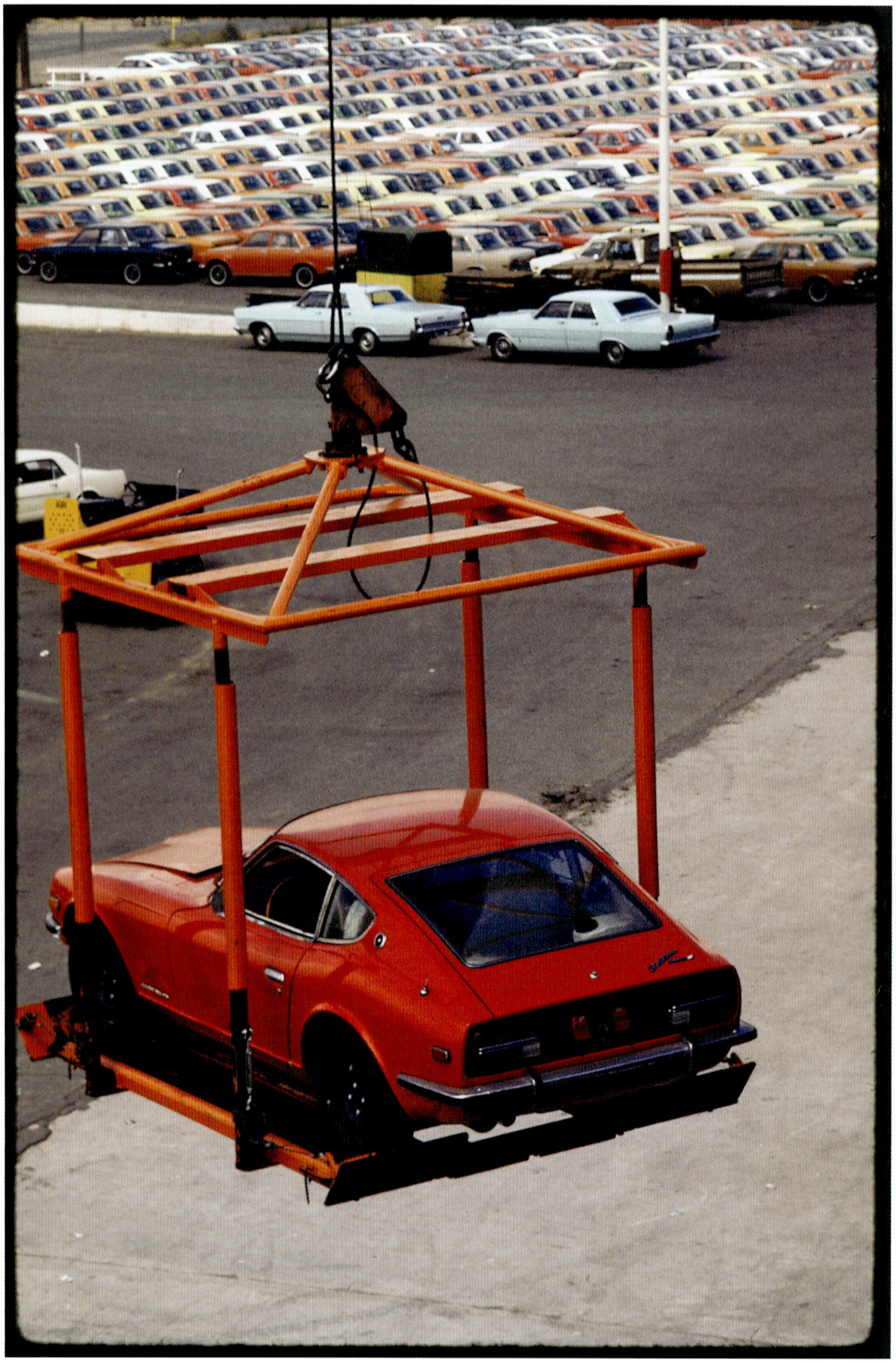

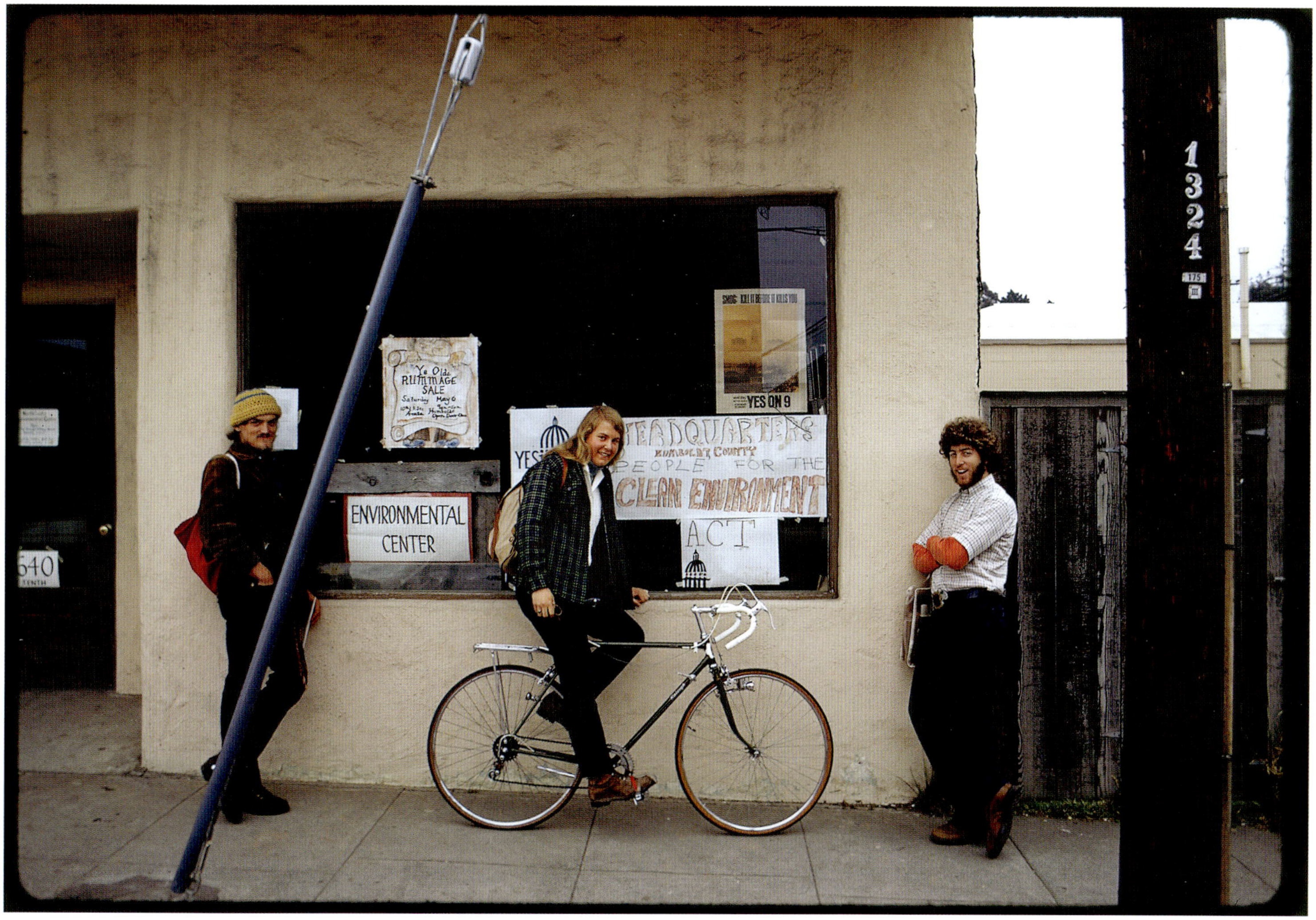

"Cyclist in front of environmental center."

Tomas Sennett, Humboldt County, California, May 1972

(412-DA-539)

Opposite: "Datsun being unloaded."

Gene Daniels, Terminal Island, California, May 1972

(412-DA-147)

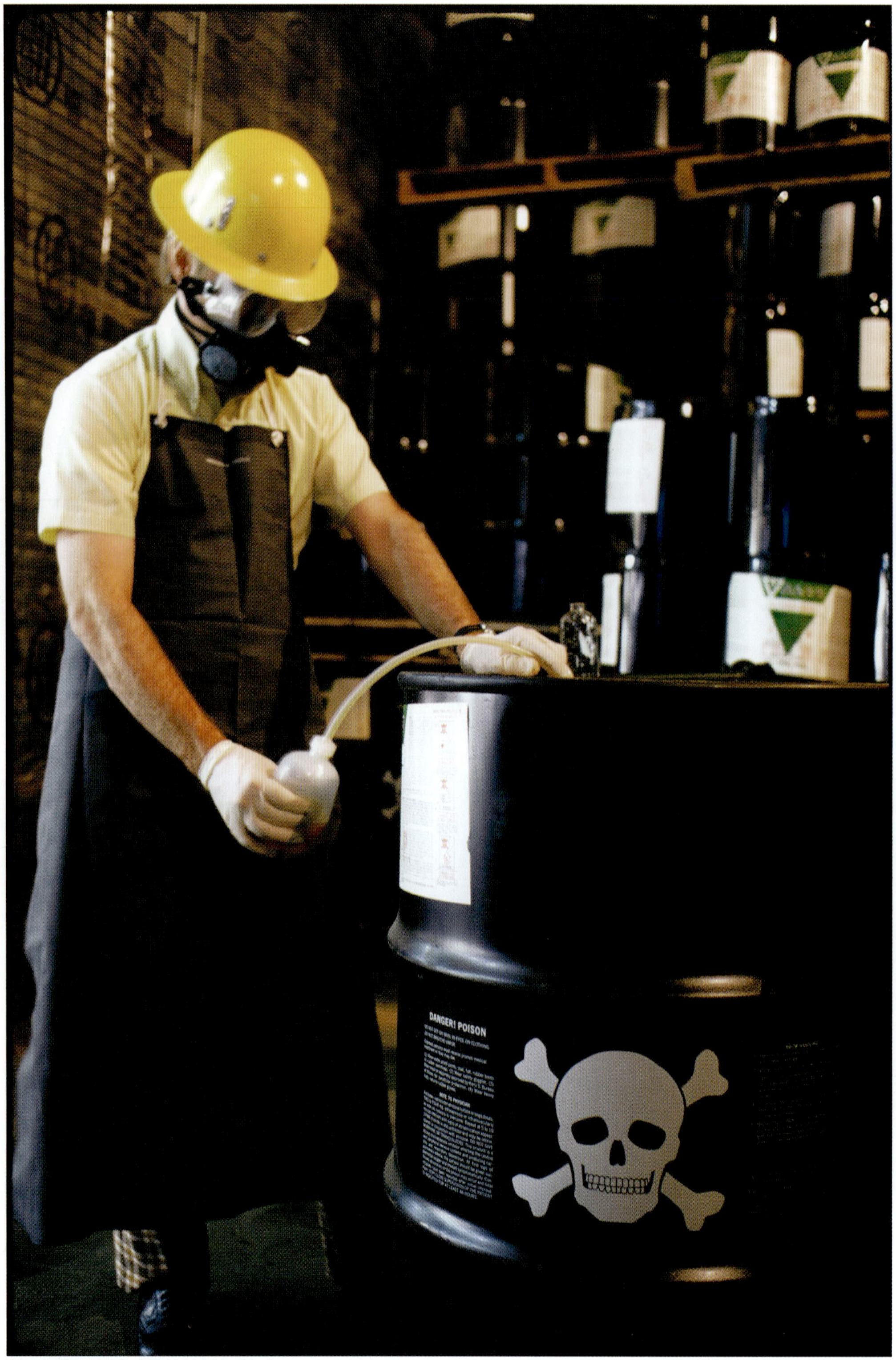
DANGER! POISON

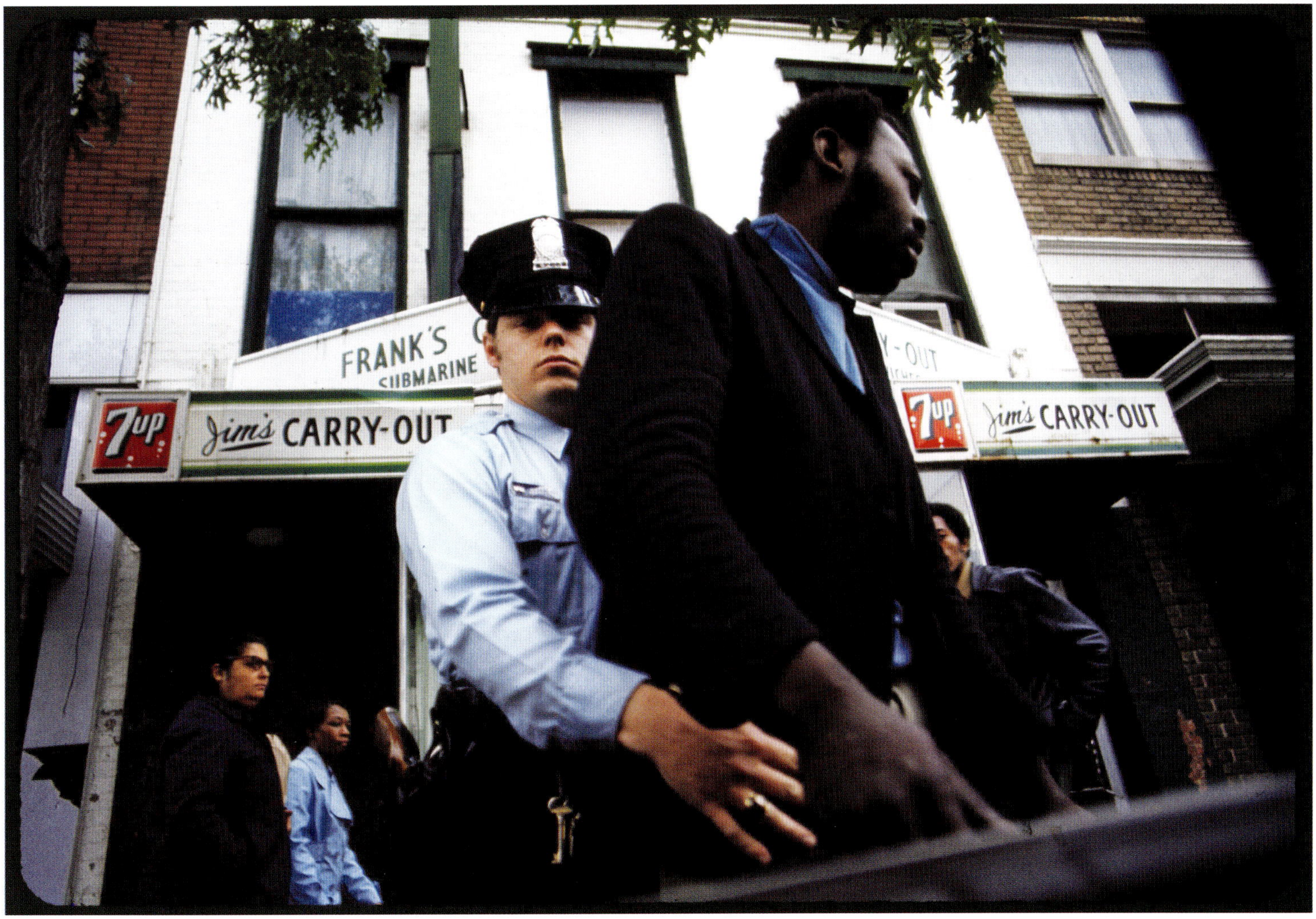

"Street arrest"

Yoichi Okamoto, Washington, DC, May 1973

(412-DA-4146)

Opposite: "An Environmental Protection Agency pesticides inspector wears a protective mask and apron as he draws a bulk sample of toxic liquid (methyl parathion) from a container during a routine inspection at the Valley Chemical Company, Greenville, Mississippi."

John Messina, Greenville, Mississippi, March 1974

(412-DA-12793)

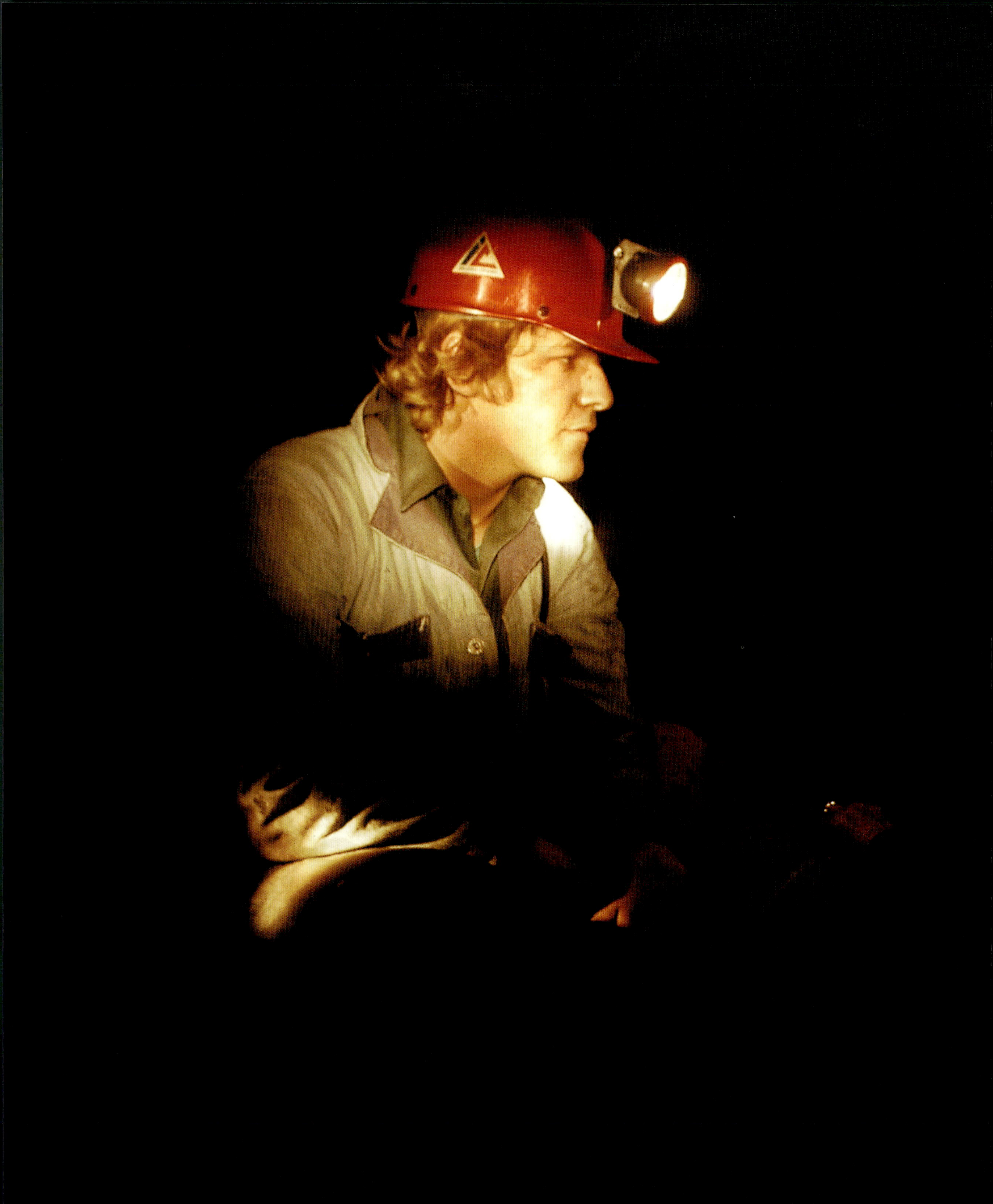

On Assignment:
Jack Corn's Appalachia

When Jack Corn took the mines, towns, and people of Appalachia as his subjects for a 1974 DOCUMERICA assignment, he was returning to topics he had been photographing since 1961. Traveling through West Virginia, Kentucky, and Virginia, Corn photographed new coal-mining technologies, Appalachia's "dismal environmental conditions," and the impact of economic change in coal country. His efforts at capturing the "lifestyles and living conditions in the mining towns" show miners changing shifts, drinking in small-town bars, enjoying their families, and receiving treatment for black lung disease. DOCUMERICA, he has written, was "one more chance to do what became a major life's work."

After DOCUMERICA, Jack Corn worked for 23 more years as a photographer and photo editor for Nashville's *Tennessean*. In 1976, he left the newspaper to become Professional Photographer in Residence at Western Kentucky University in Bowling Green. In 1984, he became the director of photography for the *Chicago Tribune*. He returned to Western Kentucky University in 1990. His photographs have appeared in the *New York Times*, *Time*, *Newsweek*, *U.S. News and World Report*, and the *Philadelphia Inquirer*. In 1995, he received the Joseph A. Sprague Memorial Award from the National Press Photographers Association.

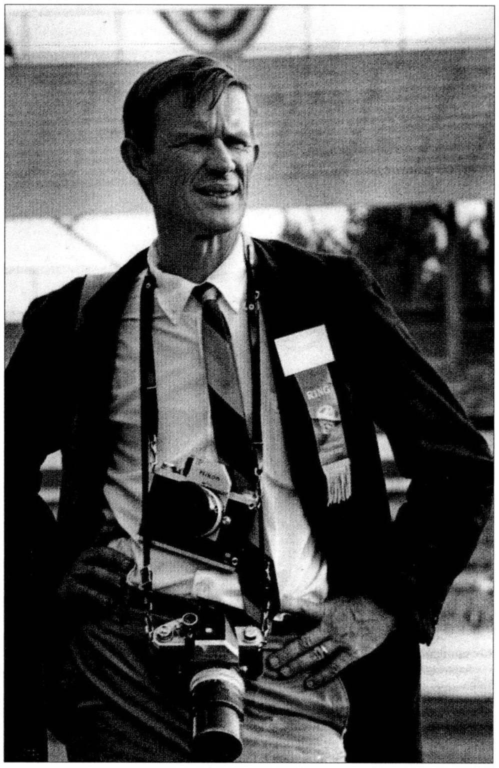

Jack Corn, 1972
Photograph by Bill Goodman
Courtesy of Jack Corn

Opposite: "Veteran miner Harold Stanley, right, talks to a young miner who has come into the mine for the first time after 40 hours of classroom training. Stanley placed his hand on the new man, shined his lamp in the miners face and said, 'Be alert, be safe, and uns (you) will be a good miner and get along just fine.' This is Virginia-Pocahontas Coal Company Mine #3, near Richlands, Virginia."

Jack Corn, Richlands, Virginia, April 1974
(detail, see page 45)

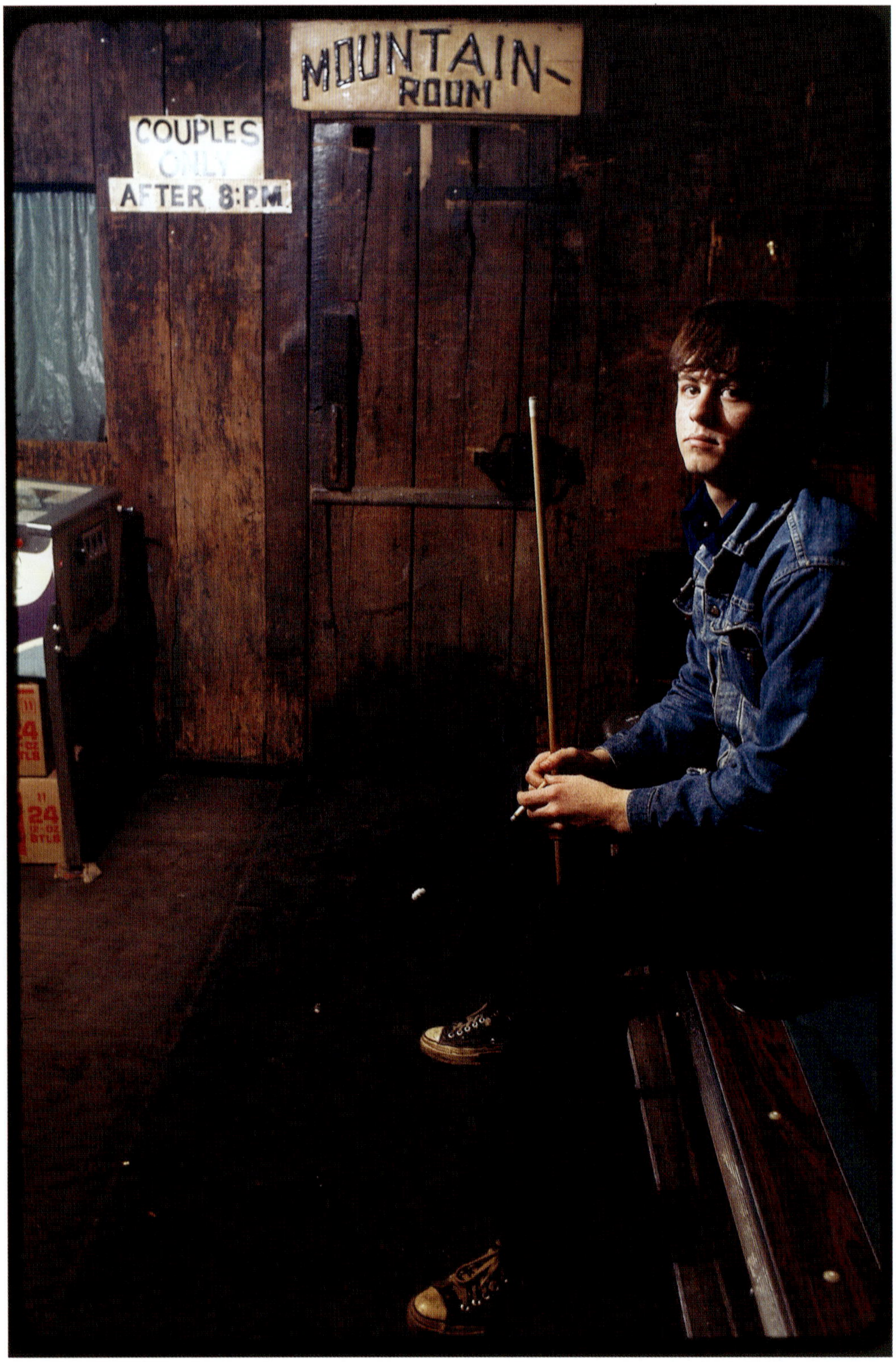
MOUNTAIN-
ROOM
COUPLES
ONLY
AFTER 8:PM

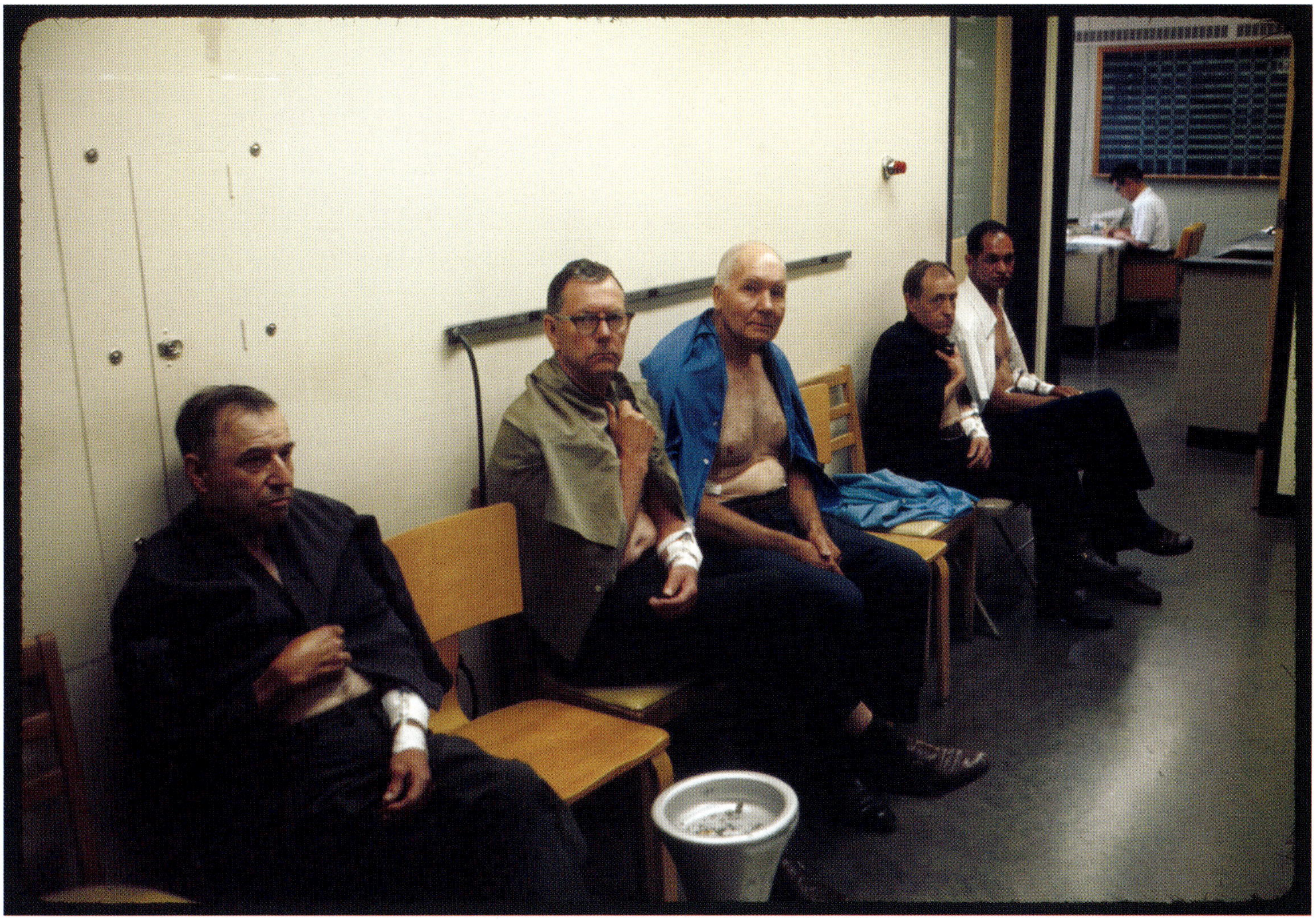

"Miners waiting for their examination at the Appalachian Regional Hospital in Beckley, West Virginia. Dr. Donald Rasmussen's black lung laboratory uses known testing methods to determine if miners are suffering from the disease which results from coal dust filling lung sacs and causing progressive shortening of breath. The lab, with funds paid mainly by the United Mine Workers union, is well known to all coal miners who come from several states."

Jack Corn, Beckley, West Virginia, June 1974

(412-DA-14102)

Opposite: "Robert Johnson, 18, sits on a pool table in a beer joint in Clothier, West Virginia, near Madison. He has passed a job physical and is waiting to go to work in the mines. Many of the young men like to get together in the taverns and drink beer and talk. There is little else to do in the small mining towns."

Jack Corn, Clothier, West Virginia, April 1974

(412-DA-13989)

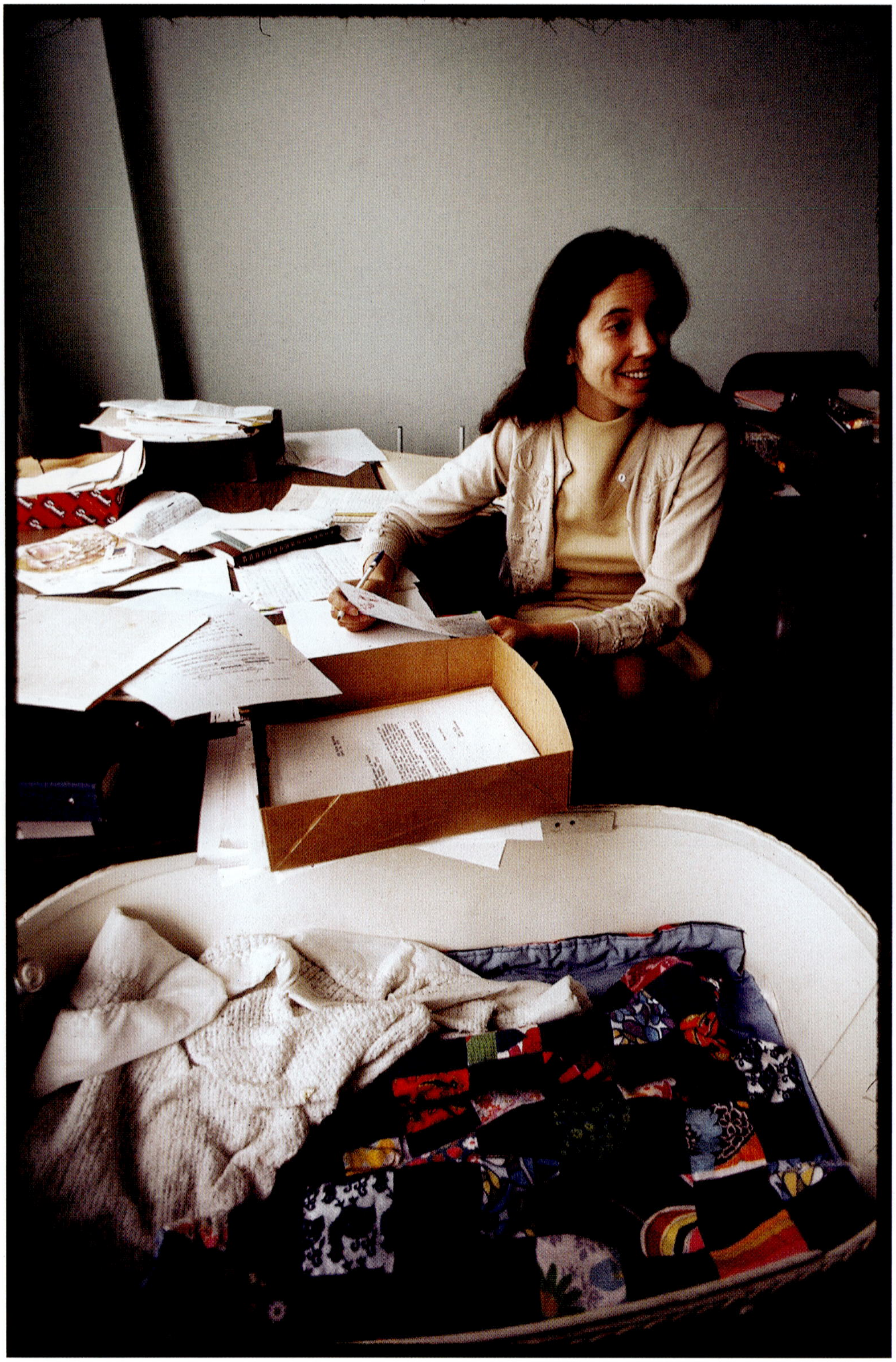

"Ms. Gail Falk Seltzer, a staff lawyer for Field Services with the United Mine Workers of America, works on Black Lung benefits at the Headquarters in Charleston, West Virginia. In the foreground is the baby bed her child sleeps in while she works. Ms. Seltzer is a Yale graduate and contributor to the equal rights amendment considered for adoption to the U.S. Constitution."

Jack Corn, Charleston, West Virginia, April 1974

(412-DA-13961)

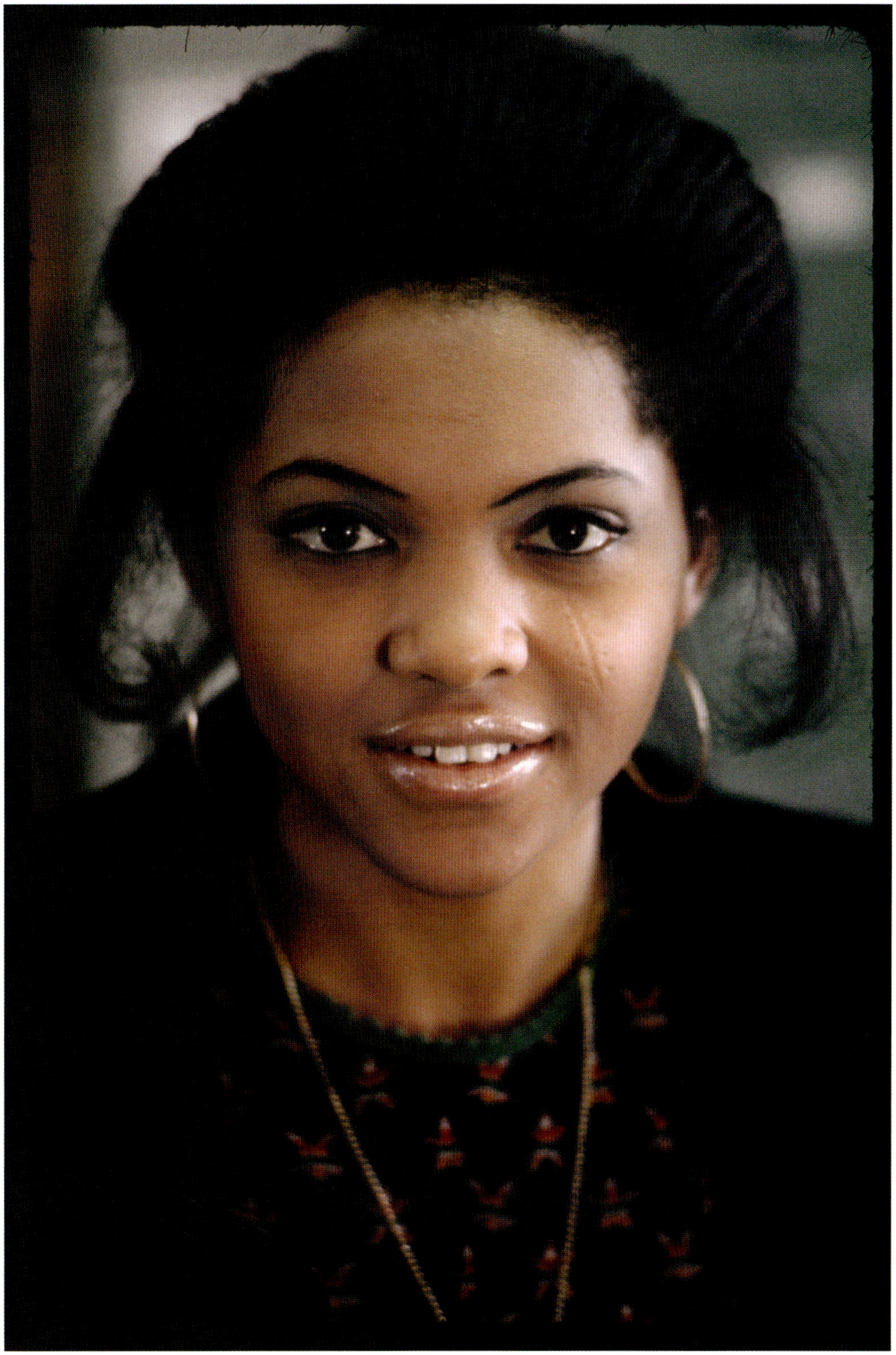

"Clarice Brown, 19, is a secretary in the United Mine Workers field Services office in Charleston, West Virginia. Her father was a miner who died of black lung disease, caused by shortage of breath as the lung sacs are filled with coal dust."

Jack Corn, Charleston, West Virginia, April 1974

(412-DA-14018)

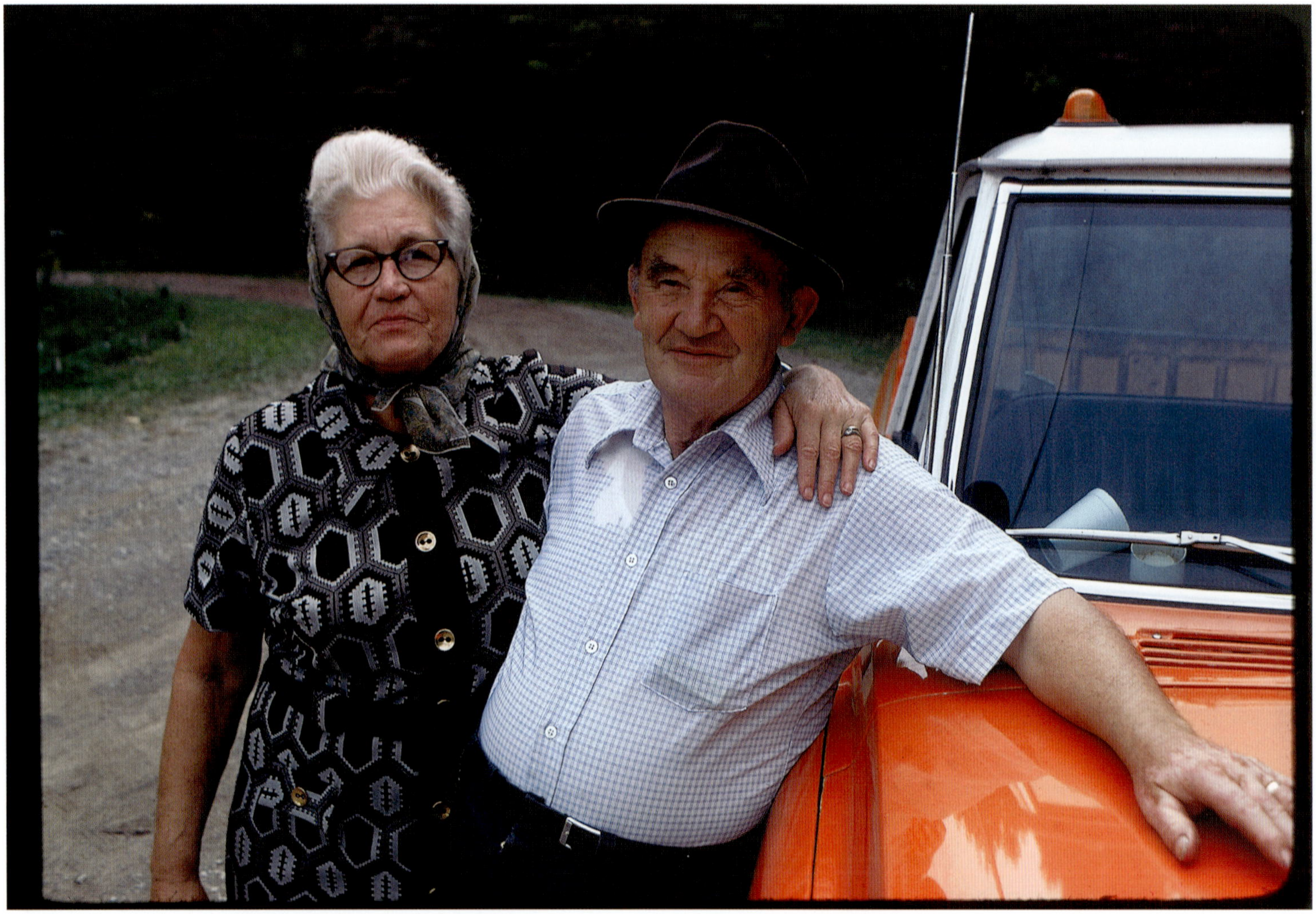

"Mr. and Mrs. Berry Howard of Cumberland, Kentucky, and the new truck he just bought with some of his black lung payments. He retired from the mines several years ago. The disease results from coal dust particles filling air sacs in the lungs and causes a progressive shortness of breath."

Jack Corn, Cumberland, Kentucky, October 1974

(412-DA-14143)

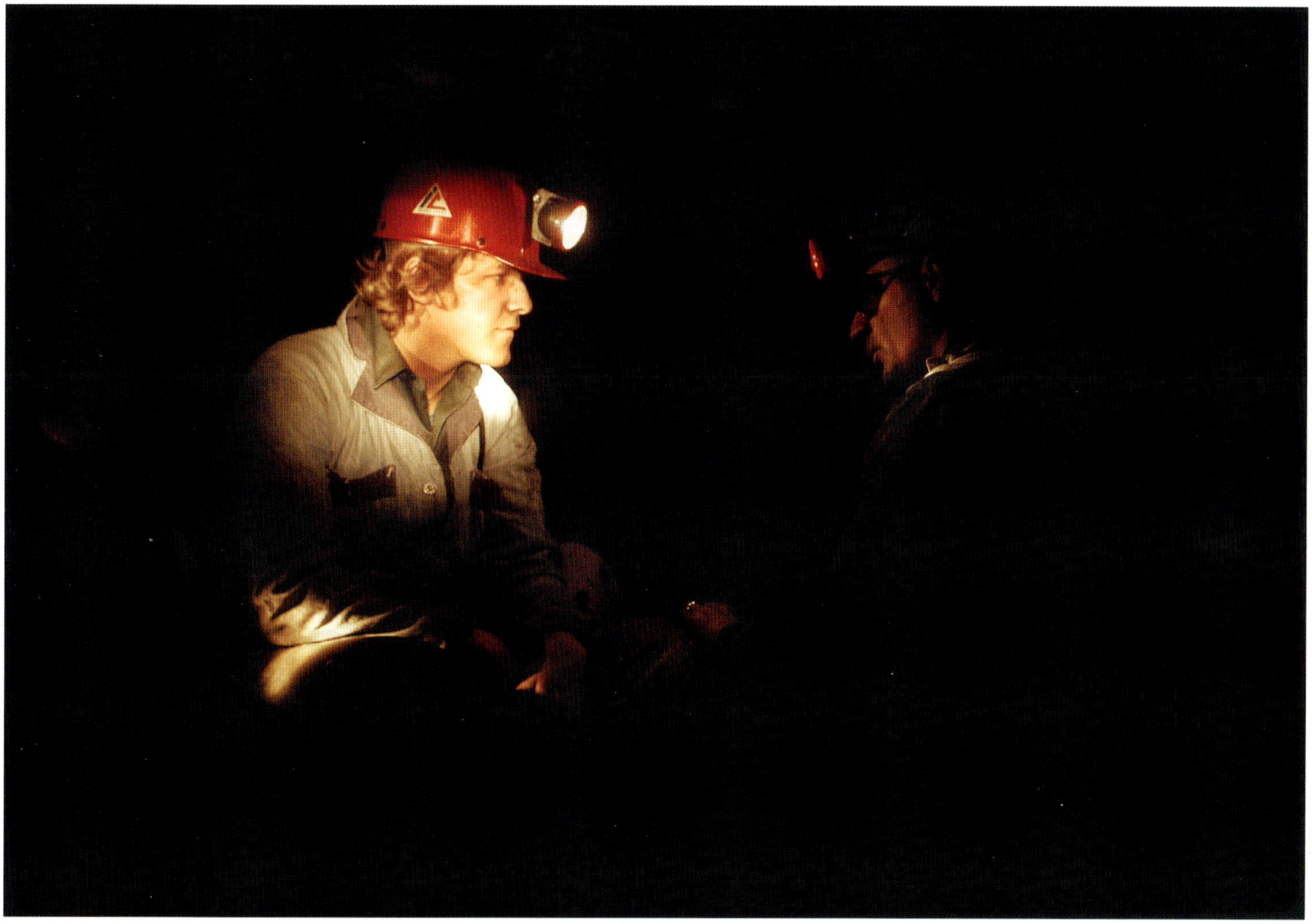

"Veteran miner Harold Stanley, right, talks to a young miner who has come into the mine for the first time after 40 hours of classroom training. Stanley placed his hand on the new man, shined his lamp in the miners face and said, 'Be alert, be safe, and uns (you) will be a good miner and get along just fine.' This is Virginia-Pocahontas Coal Company Mine #3, near Richlands, Virginia."

Jack Corn, Richlands, Virginia, April 1974

(412-DA-13910)

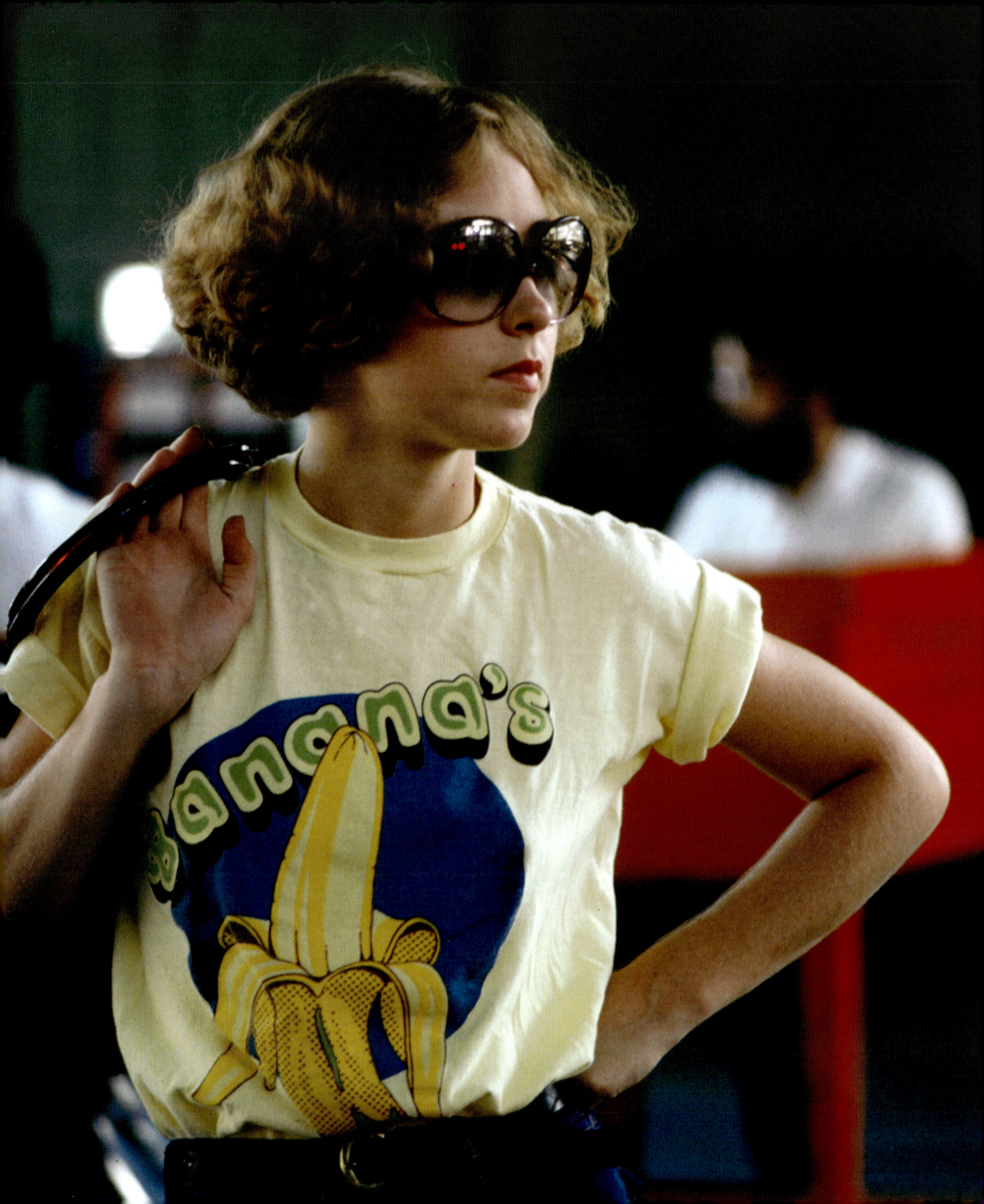
Banana's

Everybody Is a Star

"Everybody is a star
Who can rain, chase the dust away
Everybody wants to shine."
"Everybody Is a Star," Sly and the Family Stone, 1970

During the 1970s, many ordinary Americans expressed themselves more vibrantly than in the buttoned-up '50s and early '60s. In fashion, for example, the counter-cultural styles of the late 1960s spread out from college campuses and cities to Main Street and the suburbs. Men grew beards, wore their hair long, and put on colorful clothes. Women wore mini-skirts, "granny" dresses, and pantsuits.

Those fashion choices reflected other ways in which we began to embrace differences and strive for personal freedom in the 1970s. The African American civil rights movement inspired women, gays and lesbians, seniors, Hispanics, Native Americans, and the disabled to push for more inclusion in American life. Americans from a variety of cultural backgrounds discovered their history and their genealogical "roots." The United States, it was said, was no longer a "melting pot," but a multicultural "salad bowl."

DOCUMERICA's color portraits highlighted the increasing appreciation for diversity in American life and the different ways we chose to "do our own thing."

Opposite: "Young woman watches as her car goes through testing at an auto emission inspection station in Downtown Cincinnati, Ohio."

Lyntha Scott Eiler, Cincinnati, Ohio, September 1975
(detail, see page 59)

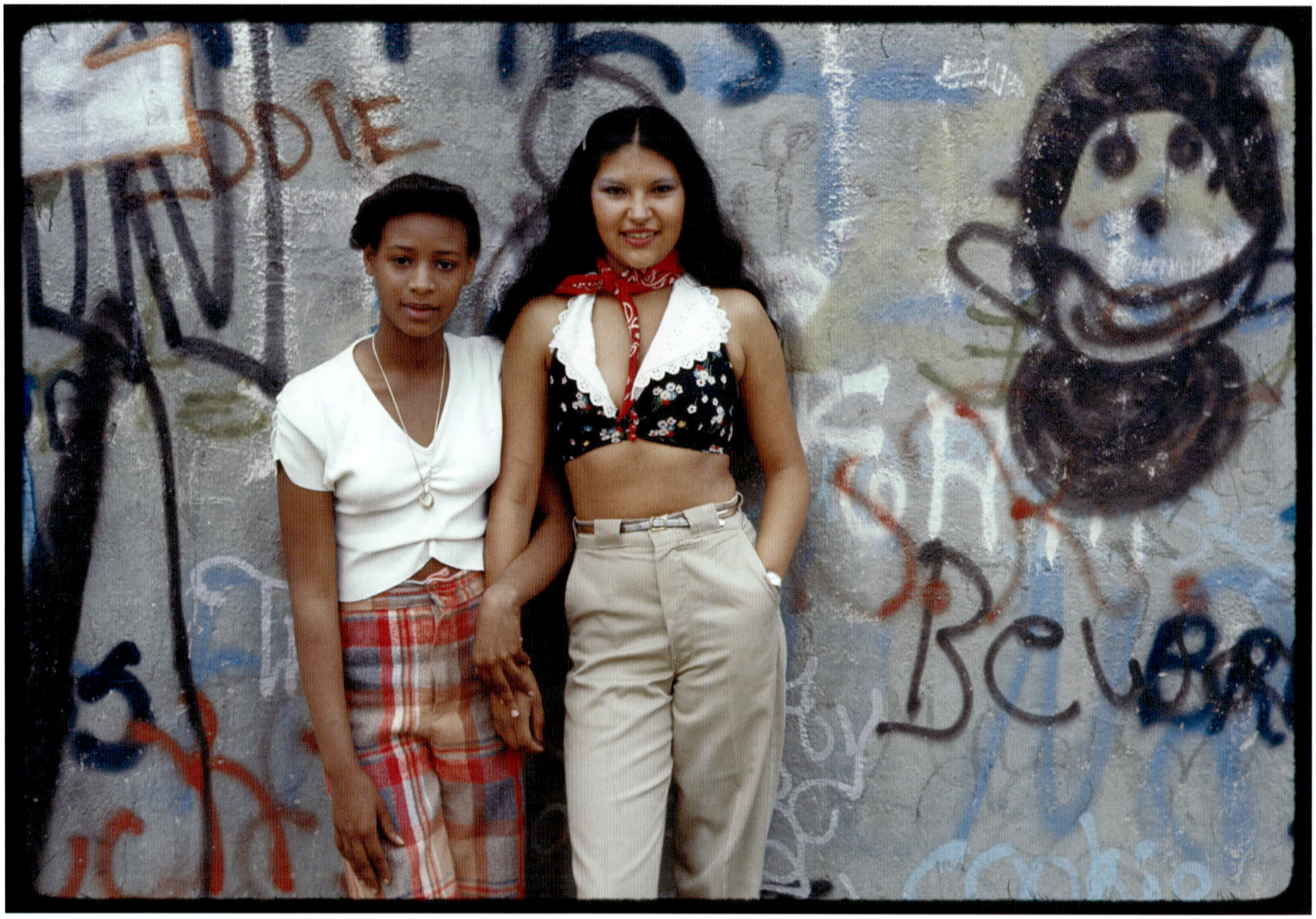

"Two Latin girls pose in front of a wall of graffiti in Lynch Park in Brooklyn, New York City. This project is a portrait of the inner city environment. It contains life, great murals on the walls of buildings and people enjoying themselves. Today's inner city is a contradiction to main stream America's gas station expressways shopping centers and tract homes. Blacks, latins and poor whites live there."

Danny Lyon, Brooklyn, New York, June 1974

(412-DA-13444)

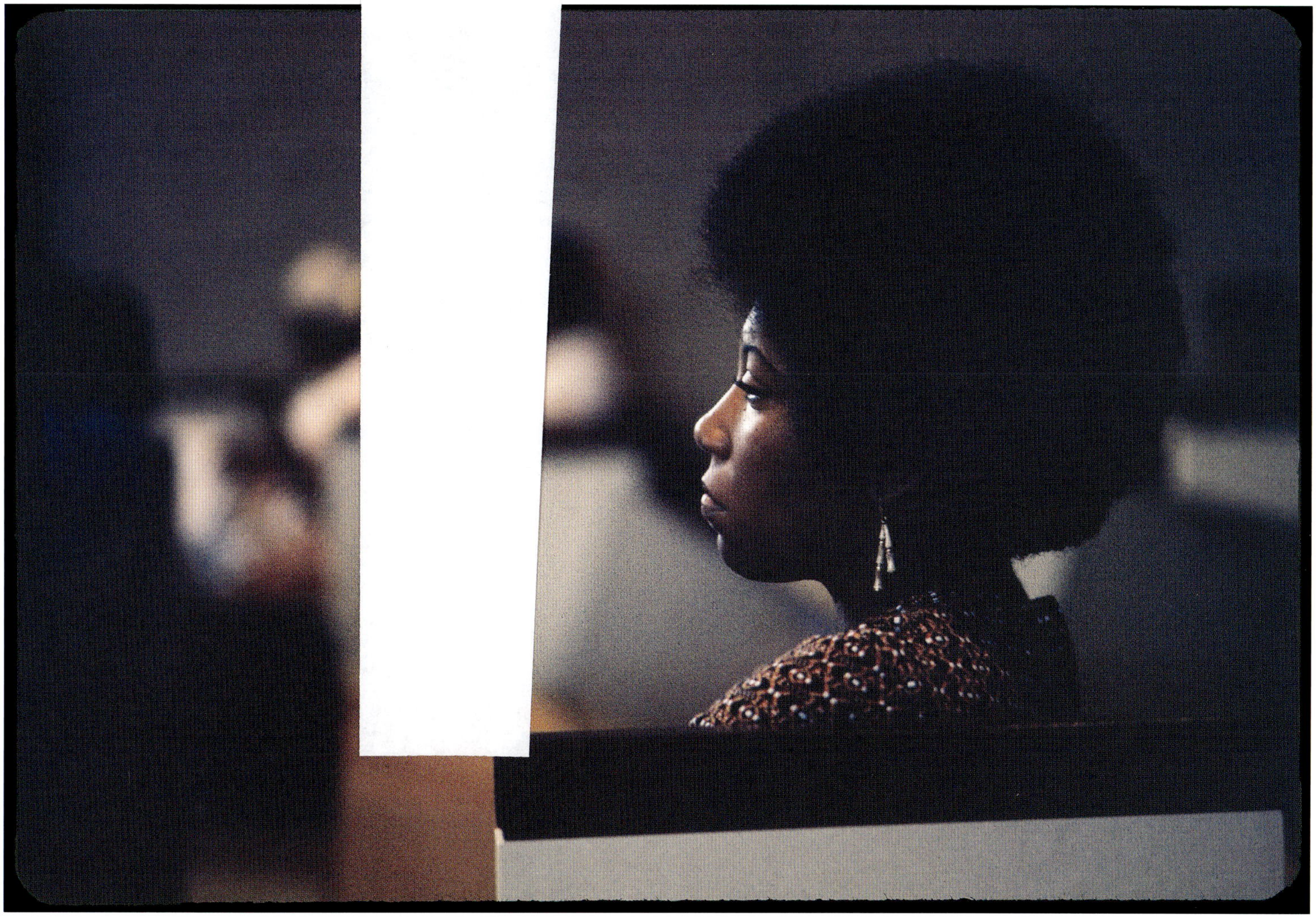

"Services at Sixth Avenue Baptist Church."

Leroy Woodson, Birmingham, Alabama, July 1972

(412-DA-2886)

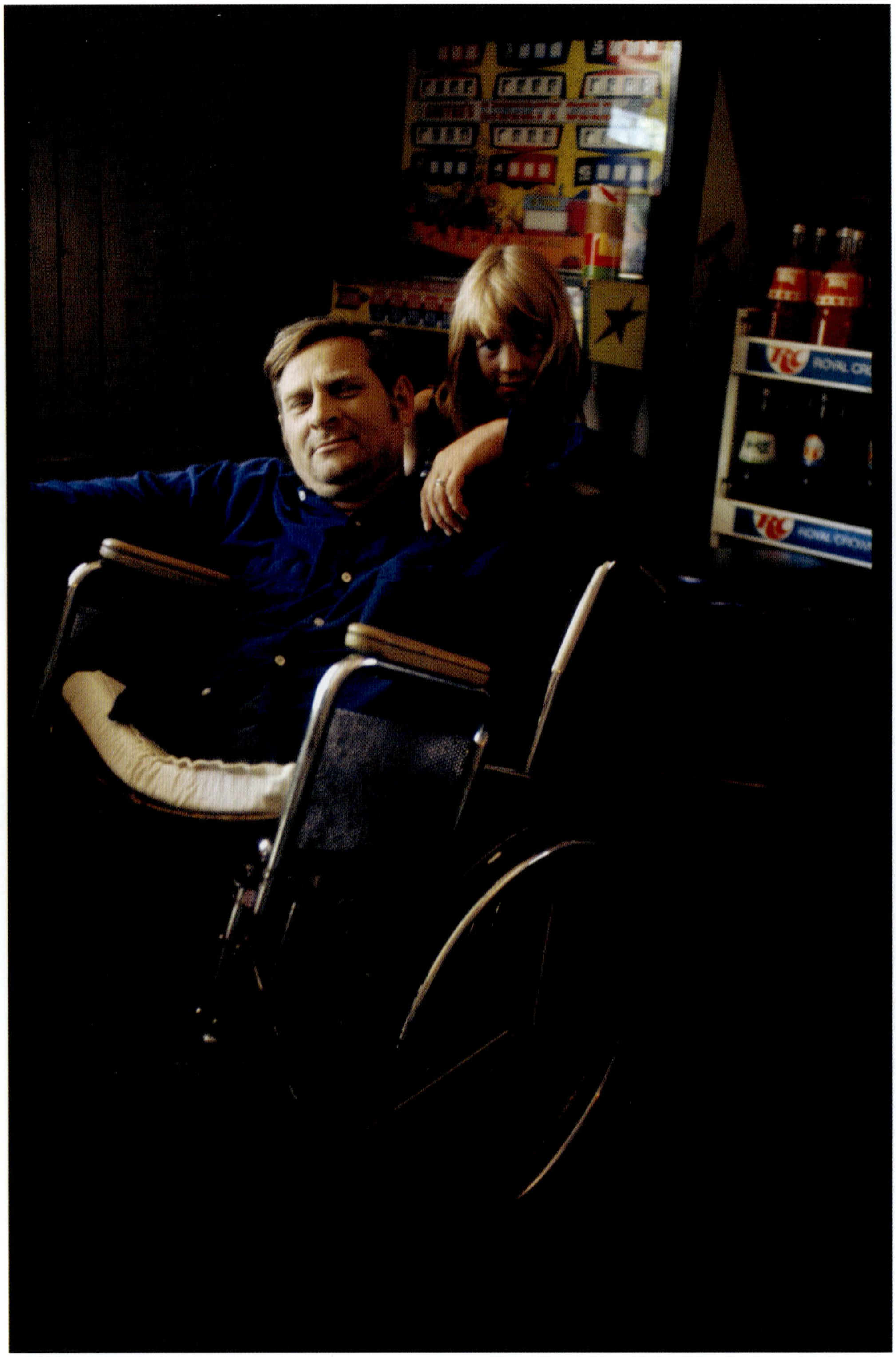
RC ROYAL CR
RC ROYAL CROWN

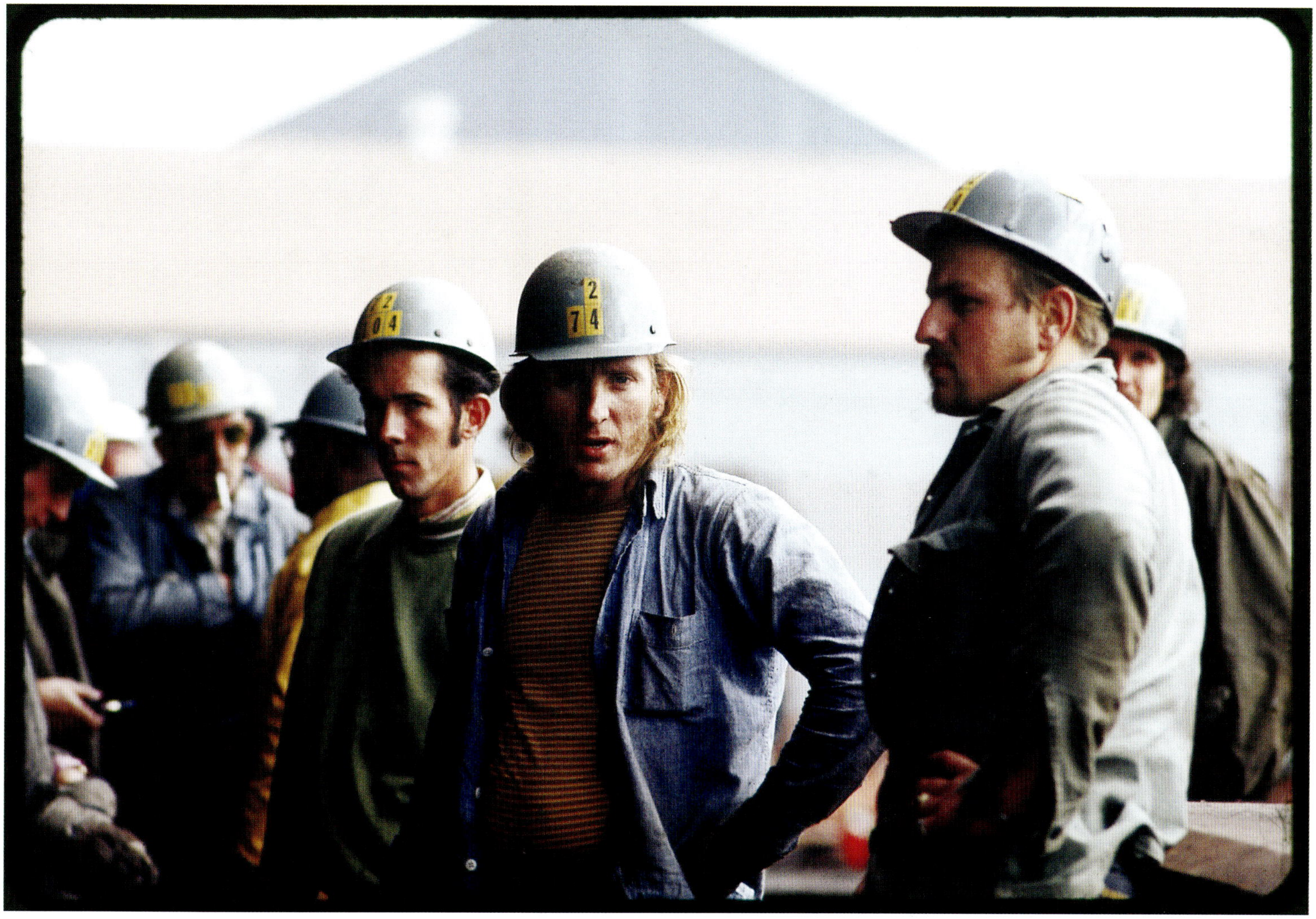

"Unloading a container ship at Dundalk Marine Terminal."

Jim Pickerell, Dundalk, Maryland, June 1973

(412-DA-4373)

Opposite: "Jack Smith, 42, a disabled miner who lives in Rhodell, West Virginia, shown with one of his daughters, Debra, in the tavern he now operates. He had worked in the mines one year when his legs were crushed in a roof cave in. It took him 18 years to receive workman's compensation. His wheelchair was bought for him by his friend, Arnold Miller, now President of the United Mine Workers. Smith is active in the union, and has manned picket lines in the past."

Jack Corn, Rhodell, West Virginia, June 1974

(412-DA-14098)

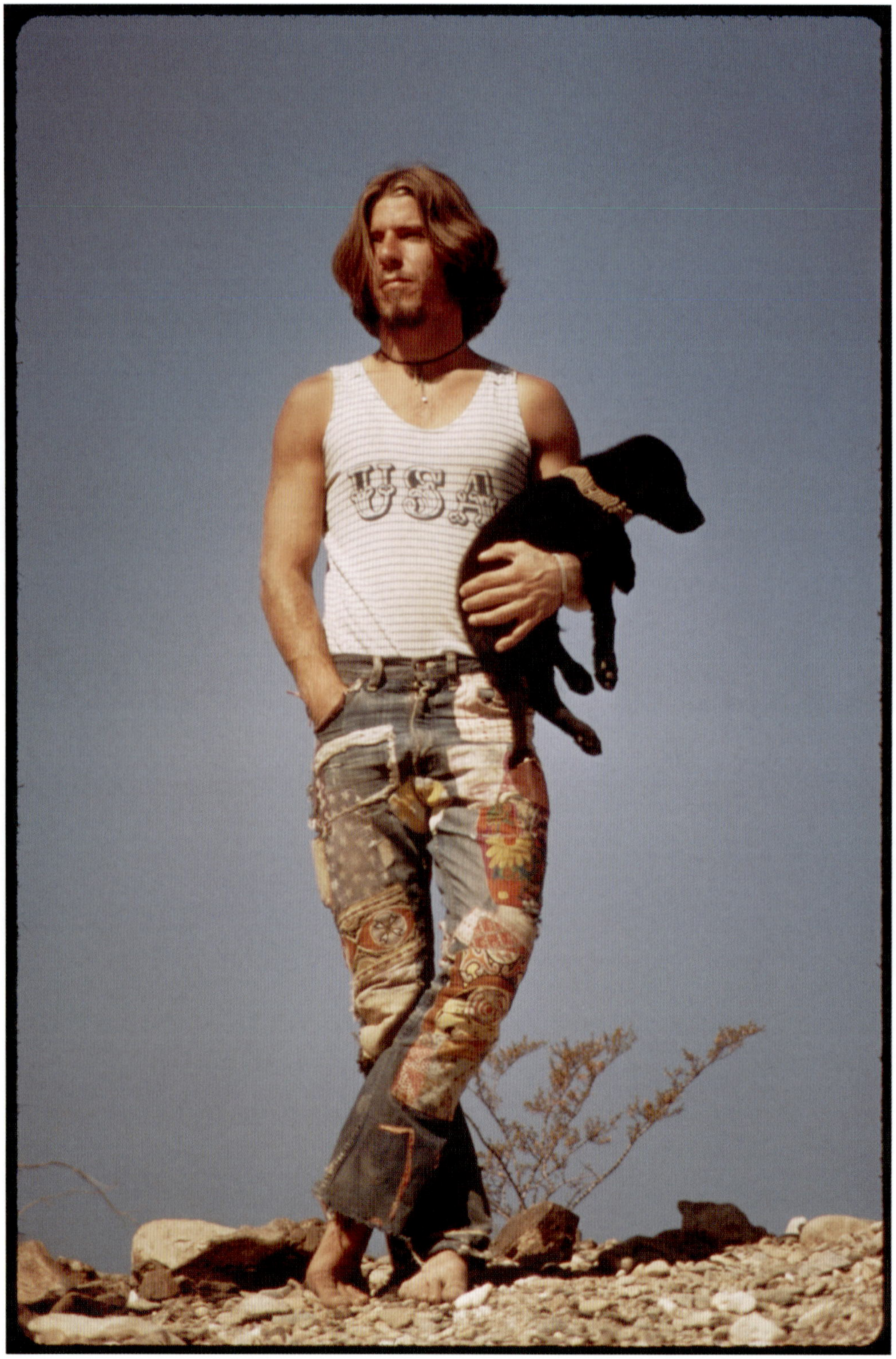

"Hitchhiker with his dog, 'Tripper,' on U.S. 66. U.S. 66 crosses the Colorado River at Topock."

Charles O'Rear, Yuma County, Arizona, May 1972

(412-DA-6626)

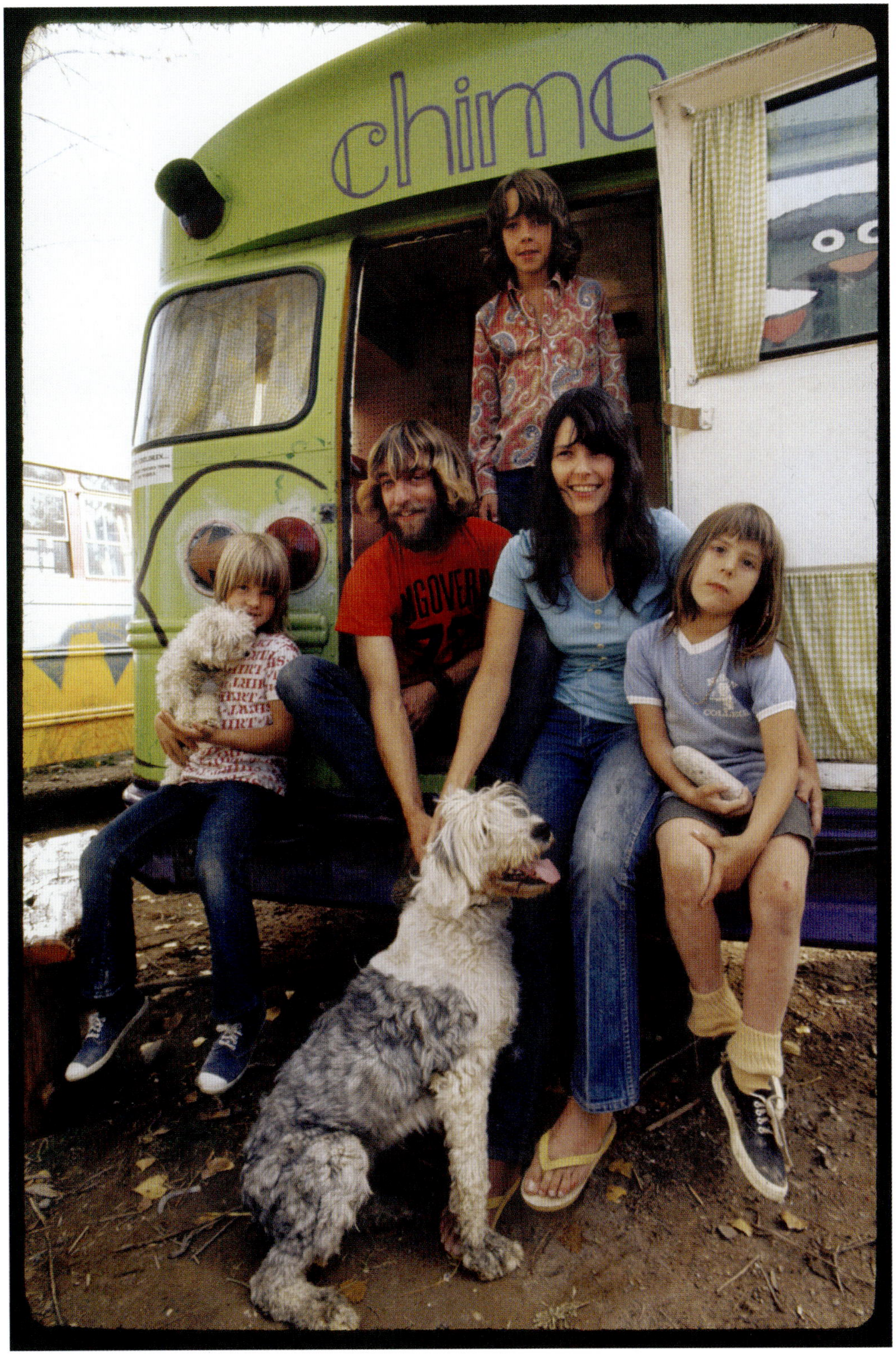

"The painted bus is home."

David Hiser, Rifle, Colorado, October 1972

(412-DA-10130)

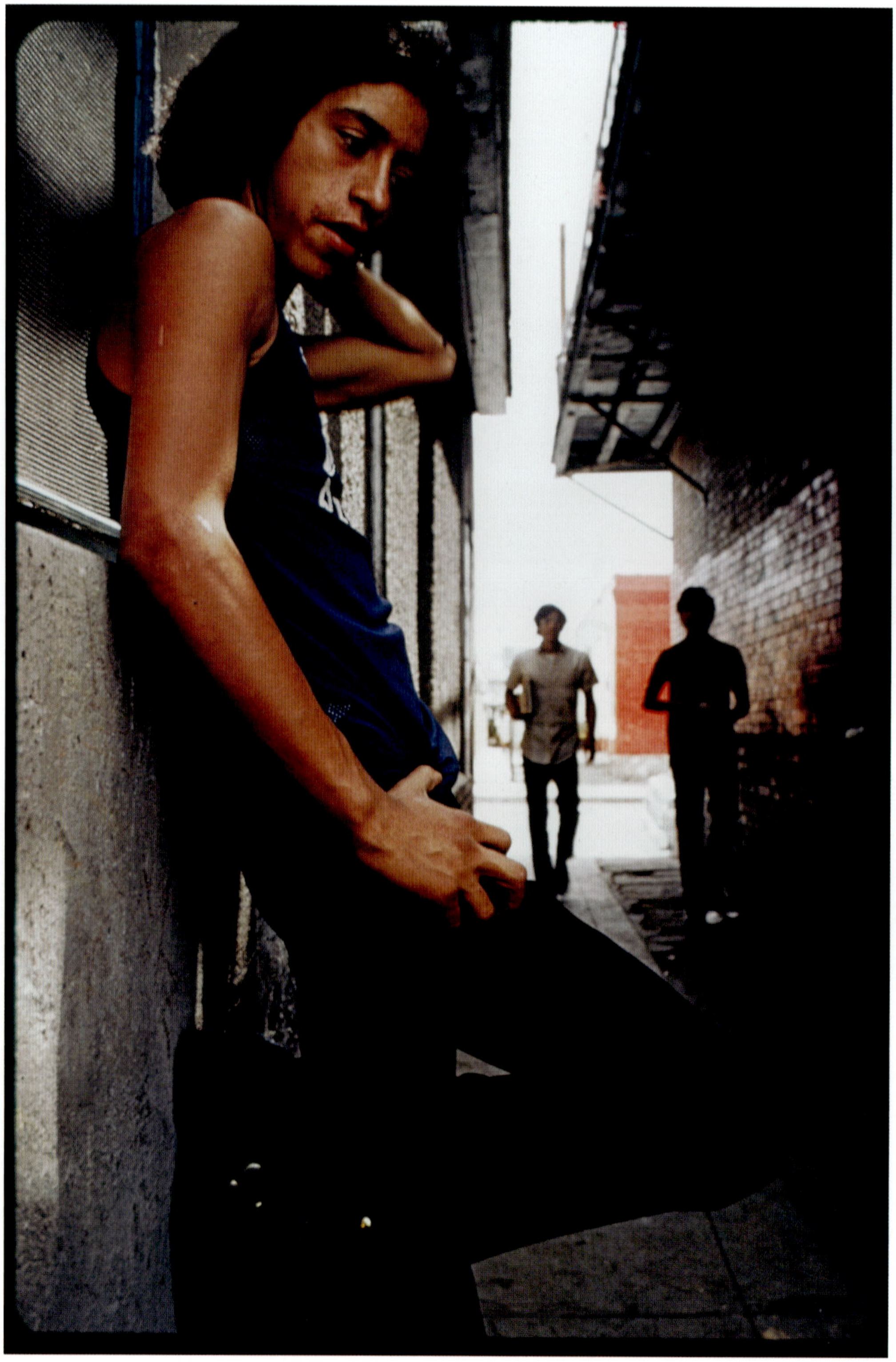

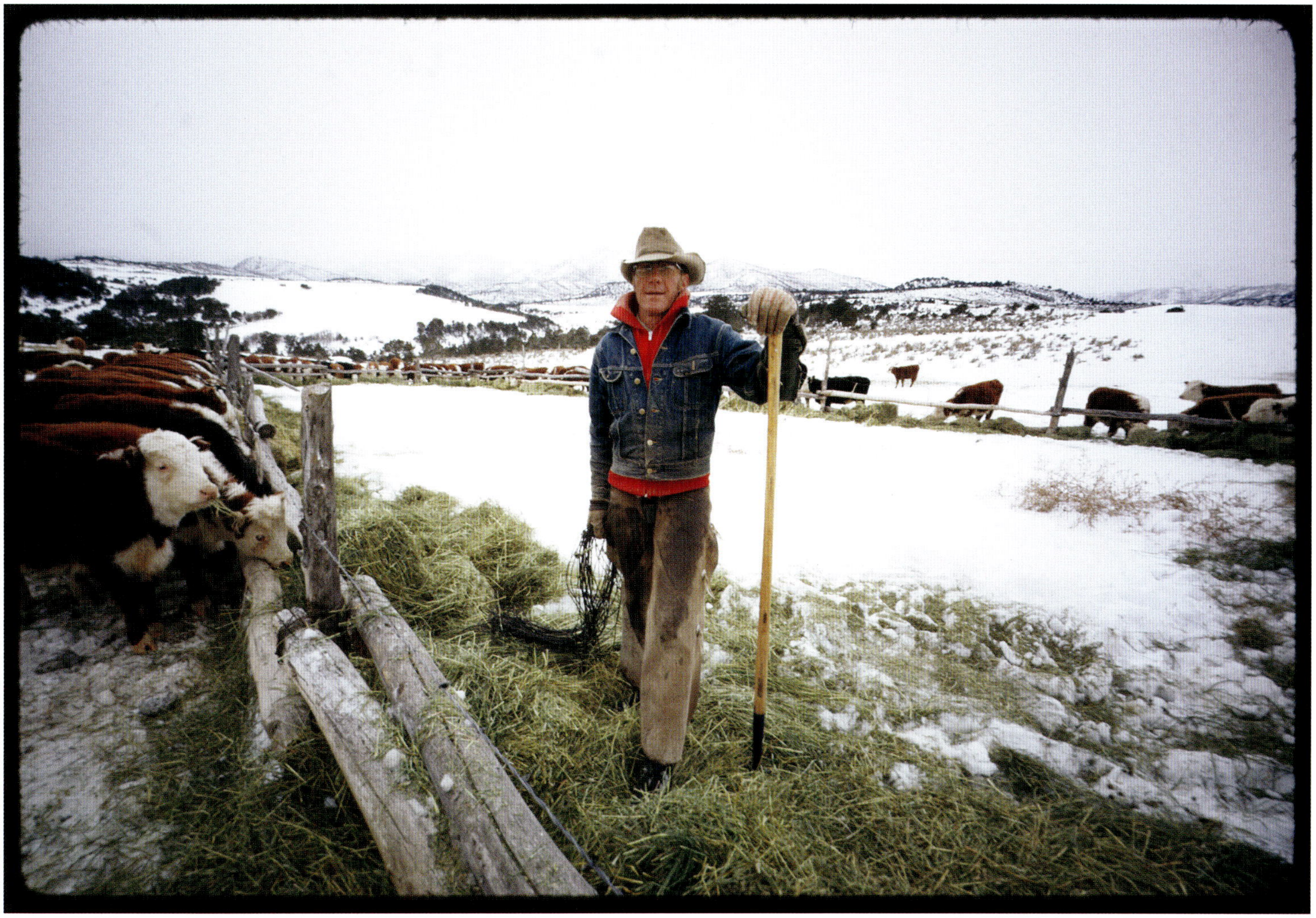

"Frank Starbuck, Last of the old time ranchers near Fairview manages a spread of 1300 acres and 400 head of cattle. He does it alone because it is too difficult and expensive to get help."

David Hiser, near Rifle, Colorado, October 1972

(412-DA-10191)

Opposite: "Chicano teenager in El Paso's Second Ward. A classic 'Barrio' which is slowly giving way to urban renewal."

Danny Lyon, El Paso, Texas, May 1972

(412-DA-2822)

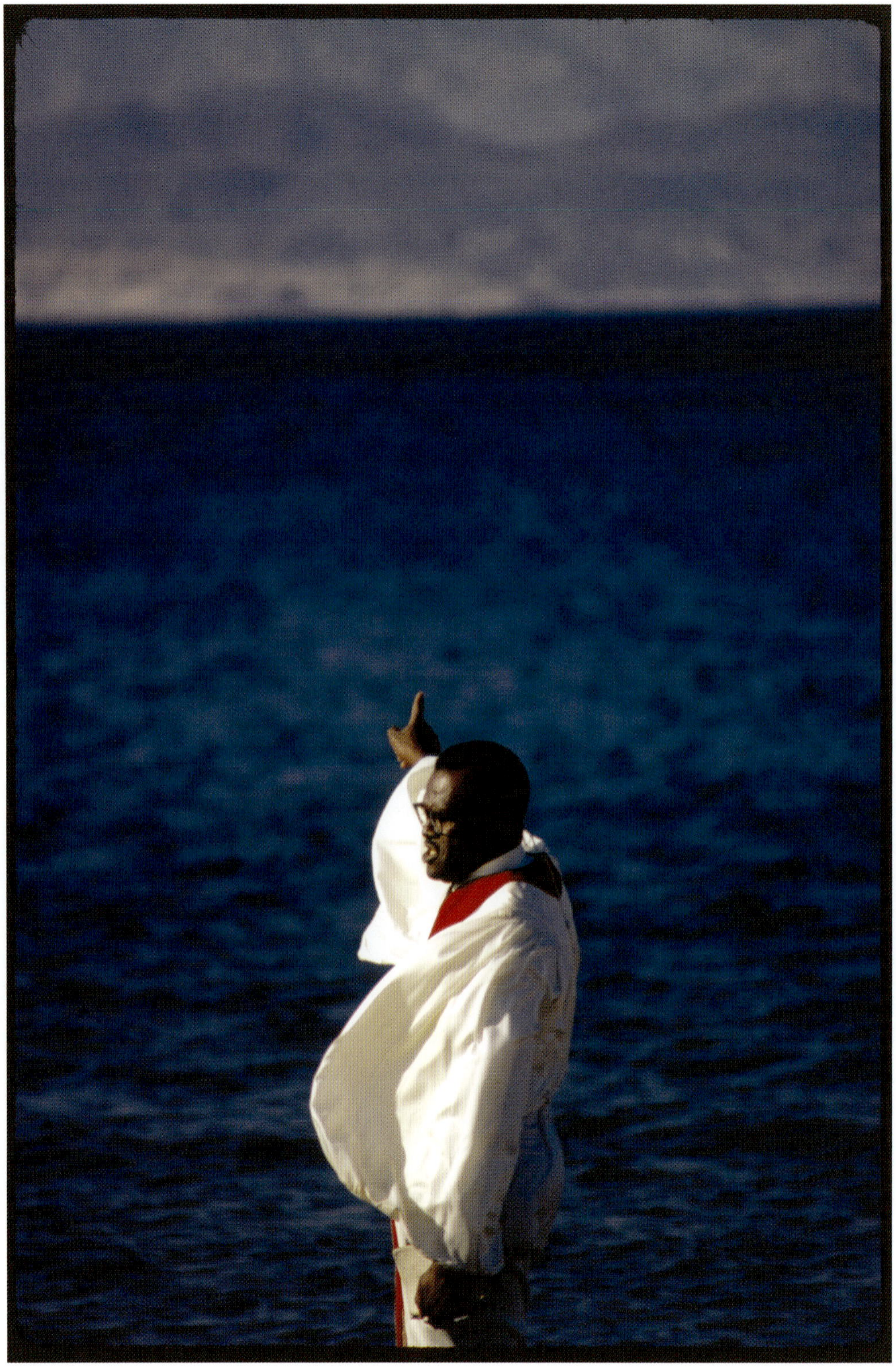

“Baptism ceremony performed by members of North Las Vegas ‘Church of God in Christ.’”

Charles O’Rear, Lake Mead, Clark County, Nevada, May 1972

(412-DA-6506)

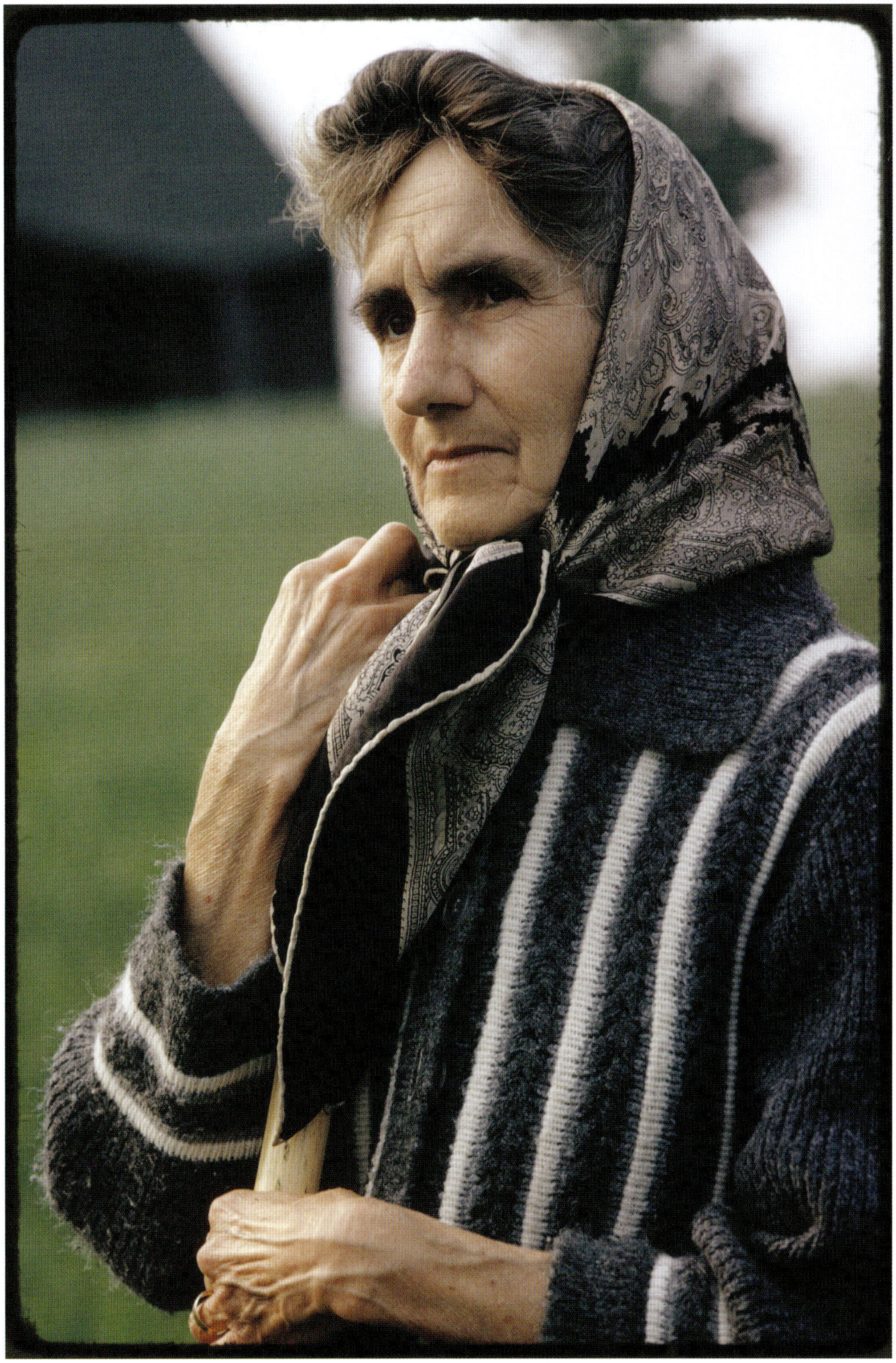

"This woman lives on a dairy farm near Randolph Center, Vermont, that has been owned by the family for six generations. Low milk prices and increasing property taxes threaten her way of life."

Jane Cooper, Randolph Center, Vermont, June 1974

(412-DA-13078)

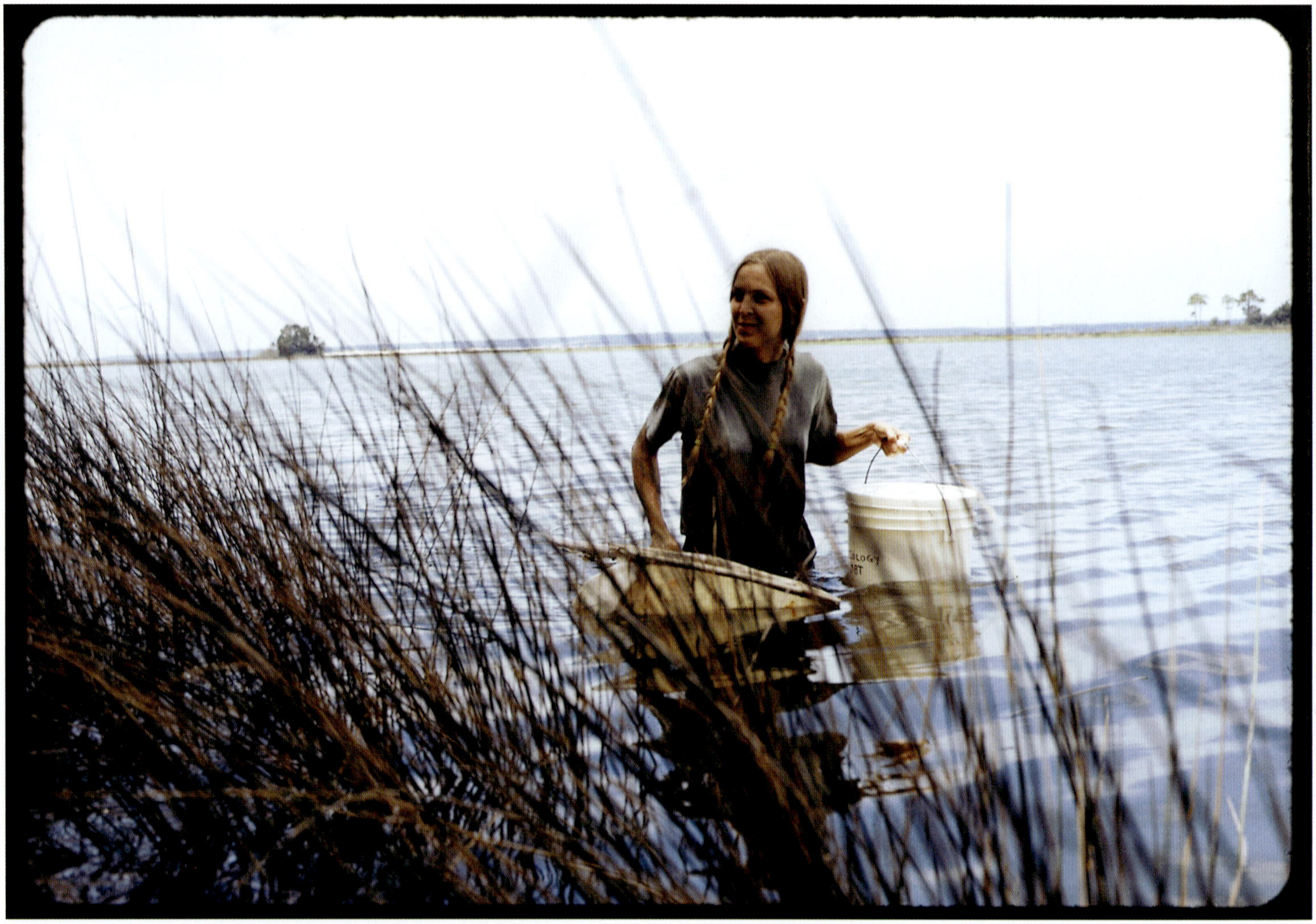

"EPA Gulf Breeze laboratory biologist is dip netting for contaminated female shrimp. This is for a study of whether PCB is passed on to their offspring."

Bill Shrout, Florida, July 1972

(412-DA-3848)

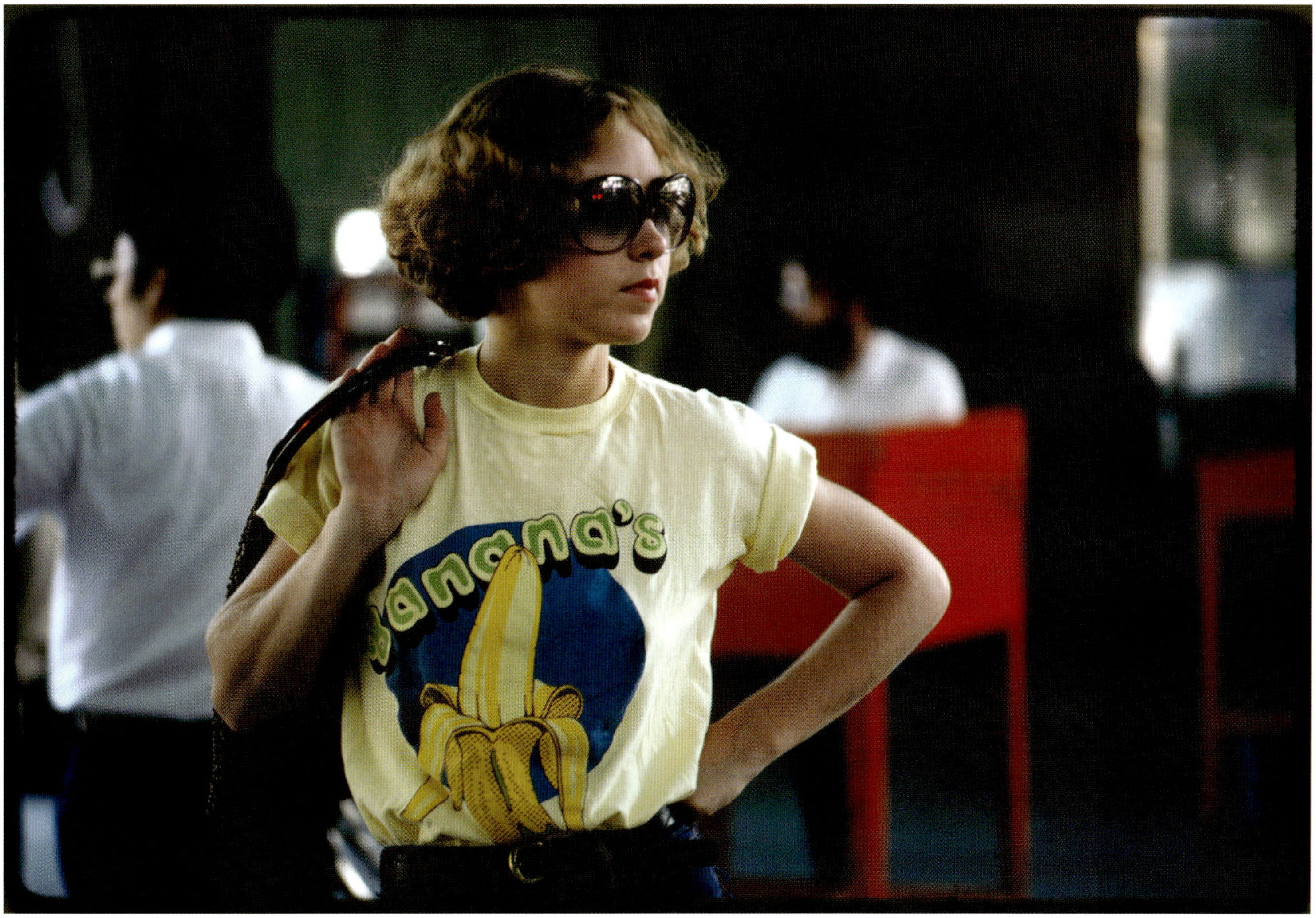

"Young woman watches as her car goes through testing at an auto emission inspection station in Downtown Cincinnati, Ohio. All light duty, spark ignition powered motor vehicles are tested annually for carbon monoxide and hydrocarbon emissions, and given a safety check. All other vehicles registered in the city receive an annual safety check. The emissions test on an exhaust analyzer went into effect in January 1975; the safety test has been in effect since 1940."

Lyntha Scott Eiler, Cincinnati, Ohio, September 1975

(412-DA-15451)

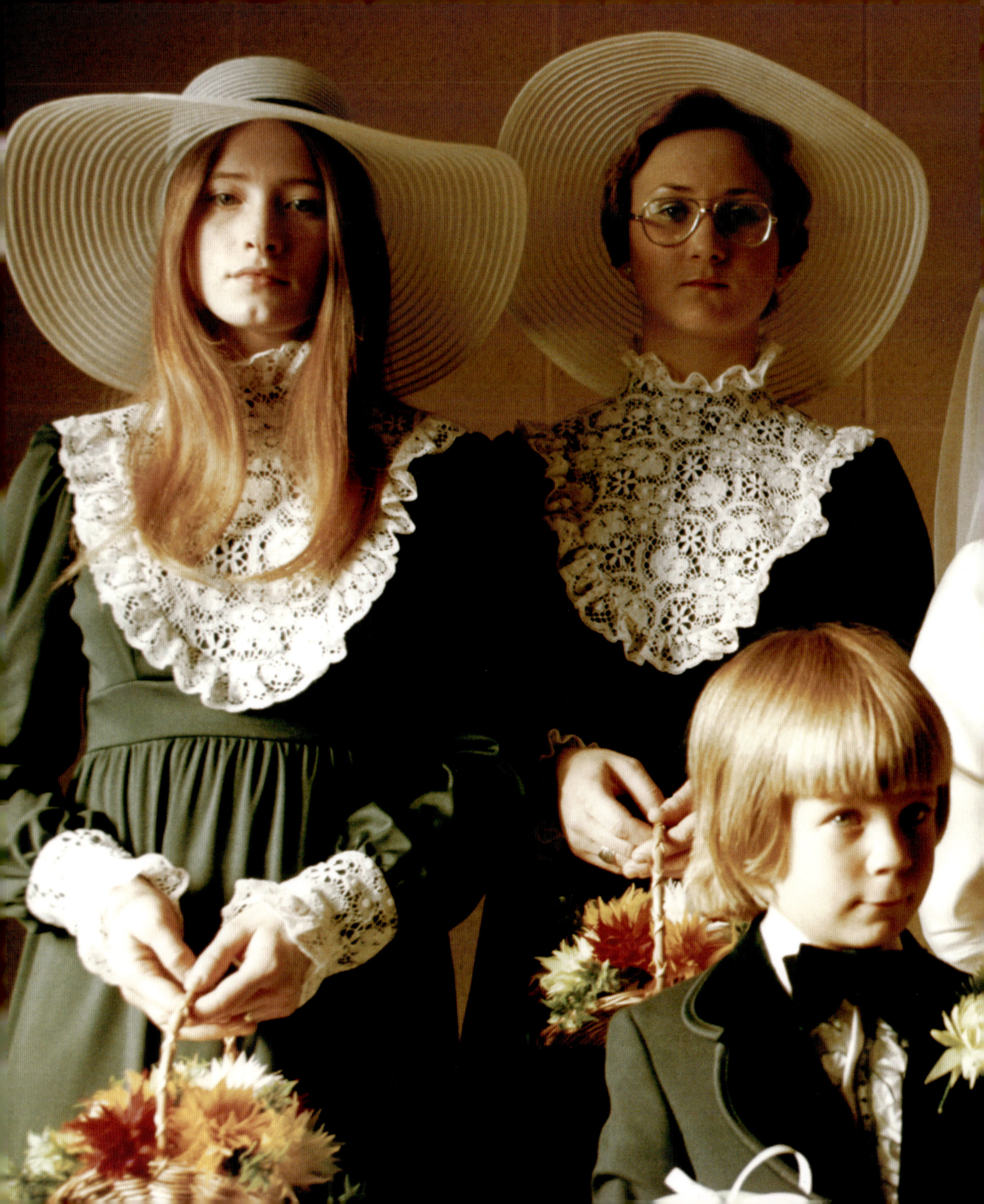

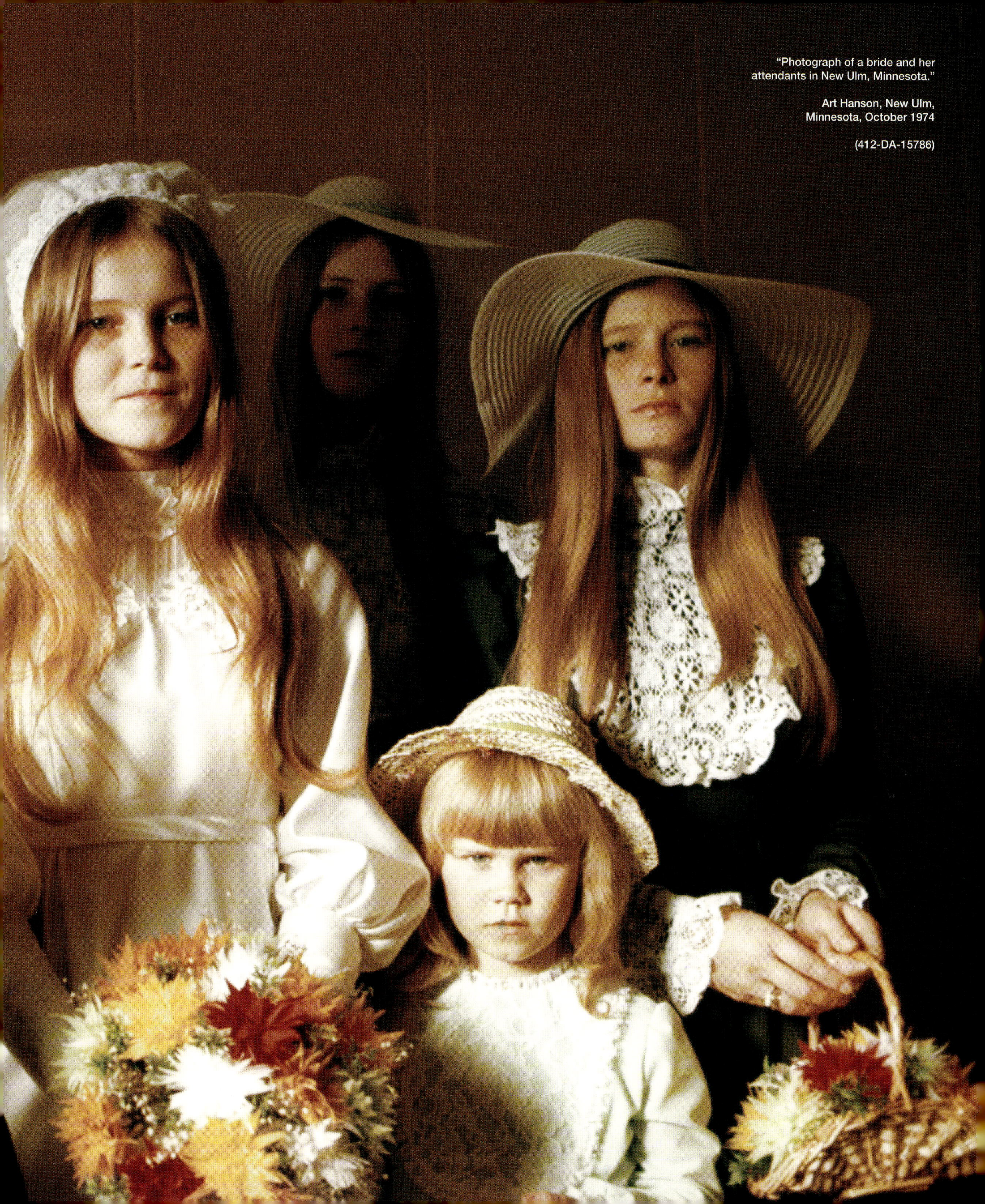

"Photograph of a bride and her attendants in New Ulm, Minnesota."

Art Hanson, New Ulm, Minnesota, October 1974

(412-DA-15786)

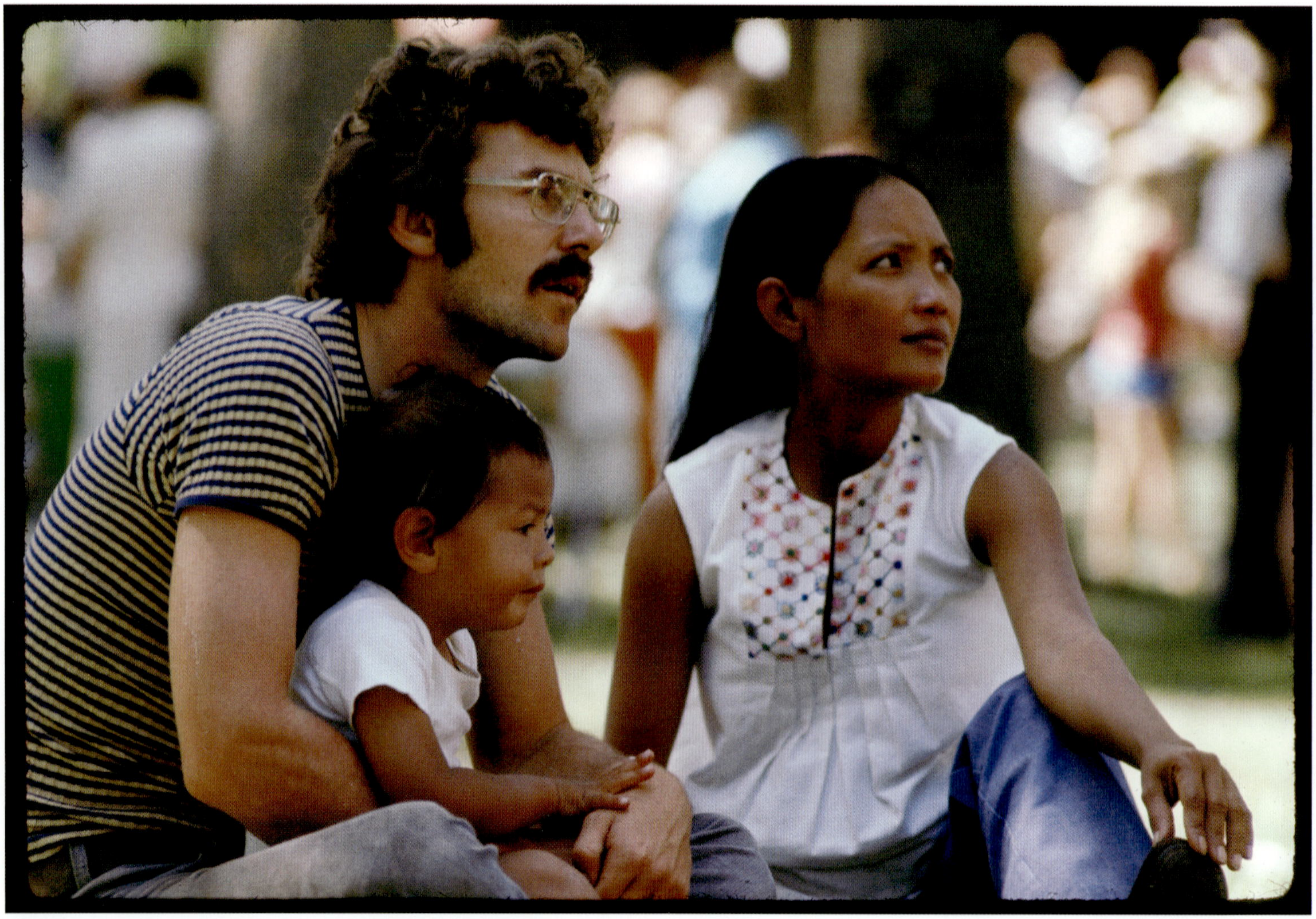

"Family enjoys a Fourth of July holiday in a park at Sleepy Eye, Minnesota, which is 12 miles west of New Ulm. Brown County is predominately a farming area where the farms are being consolidated. In the past each family owned land. Now the farms are growing larger and they are operated by two or three generations of the same family. The county was settled in the mid-1800's by Germans, Norwegians and Swedes who are clustered in various communities."

Flip Schulke, Sleepy Eye, Minnesota, July 1974

(412-DA-15737)

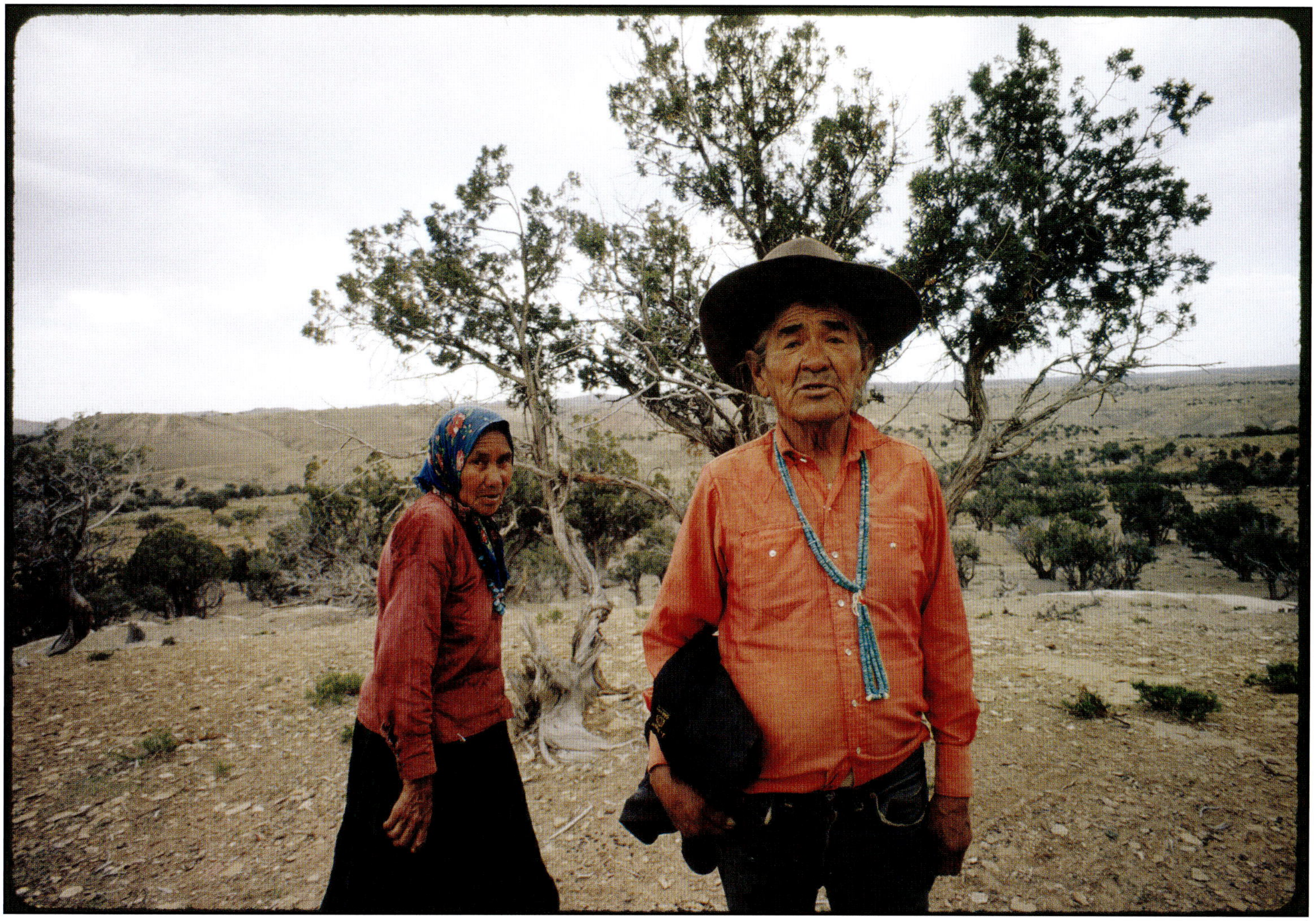

"Arizona—Navajo nation."

Terry Eiler, May 1972

(412-DA-1911)

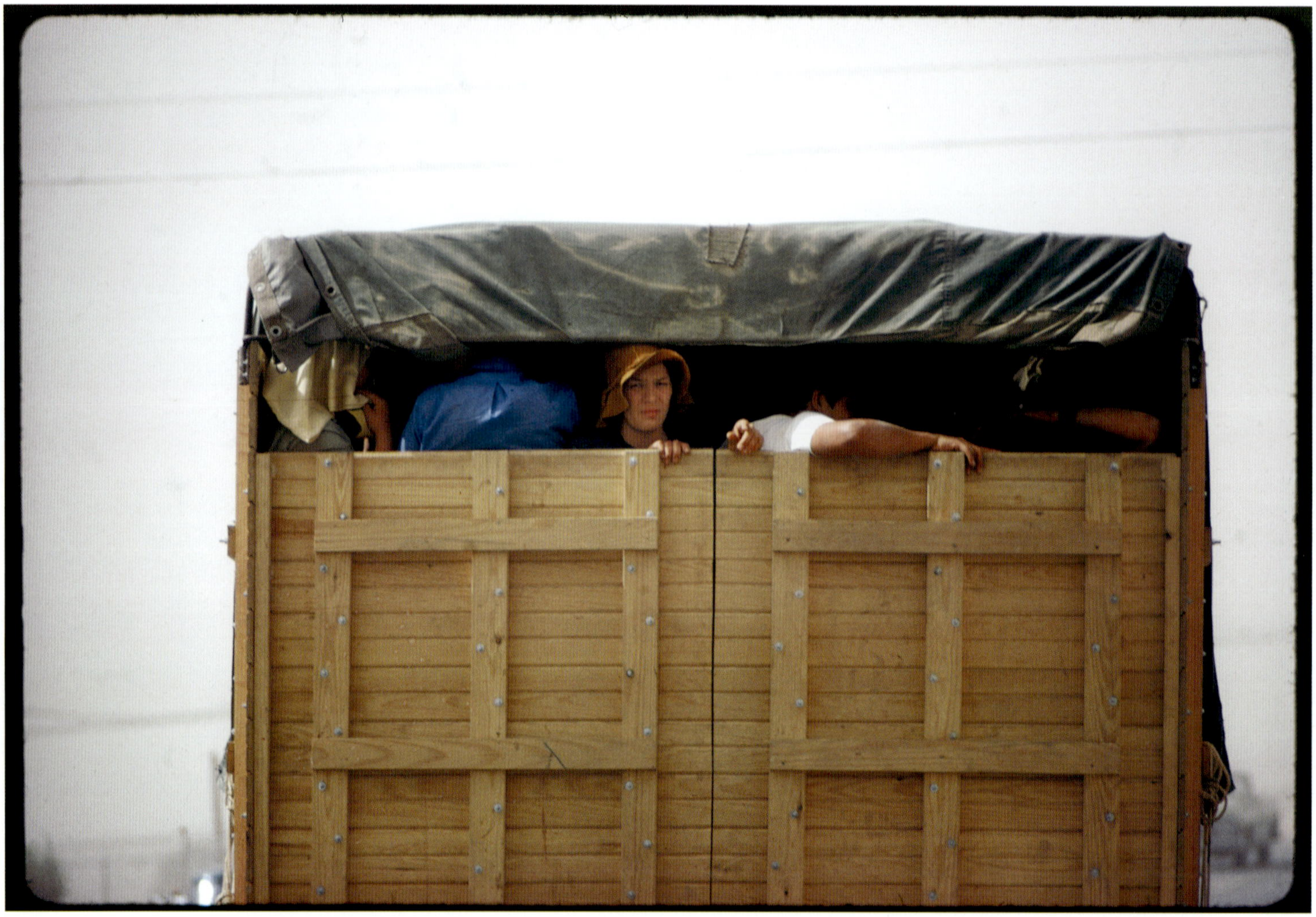

"Taking shelter during a dust storm."

Terry Eiler, Arizona, May 1972

(412-DA-11360)

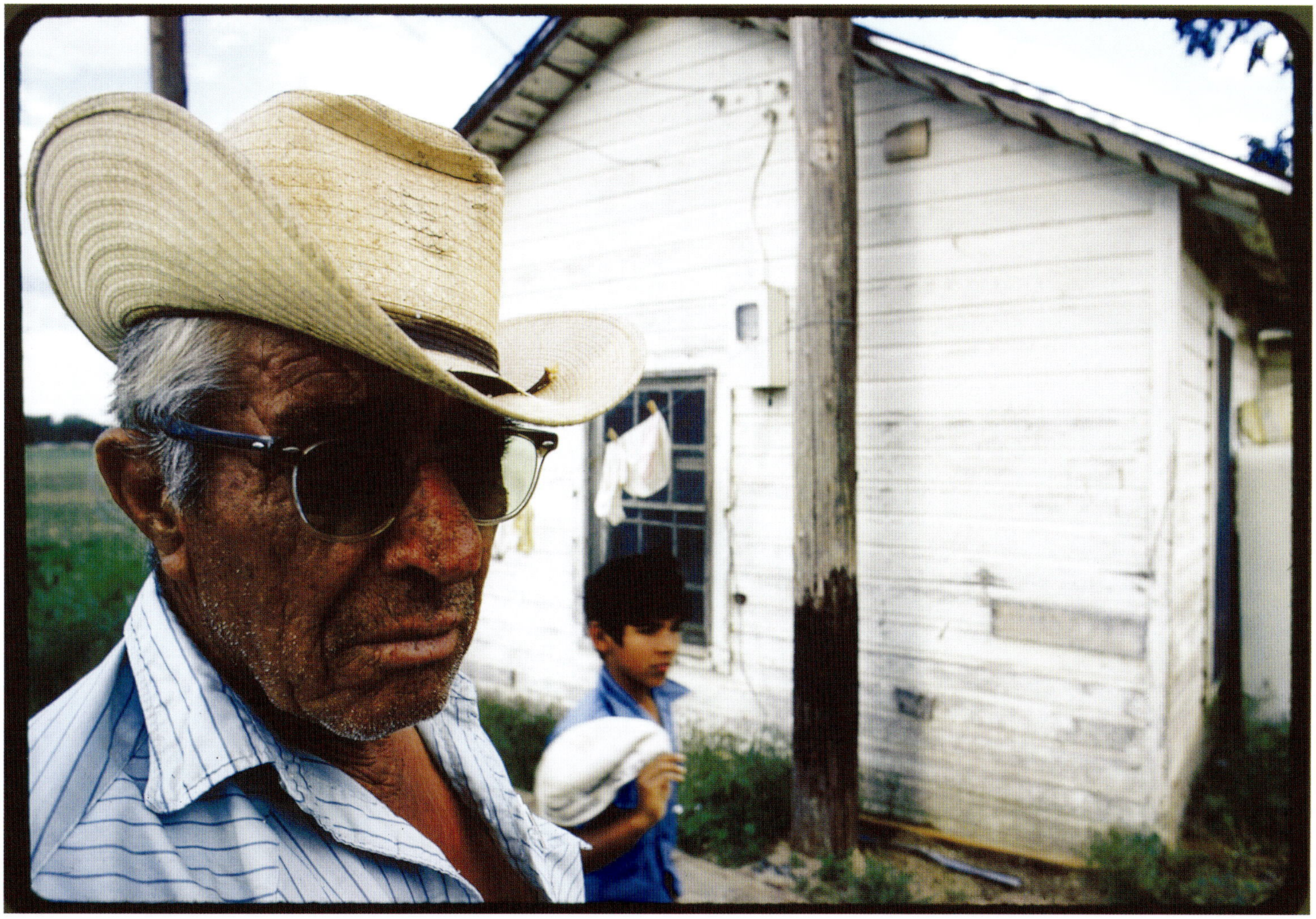

"Migrant worker with his grandchildren in front of two room shack which houses three families (20 people). This man and his family follow the crops north from Texas each year. His present job is weeding sugar beets at $2.00 an hour."

Bill Gillette, Fort Collins, Colorado, June 1972

(412-DA-1356)

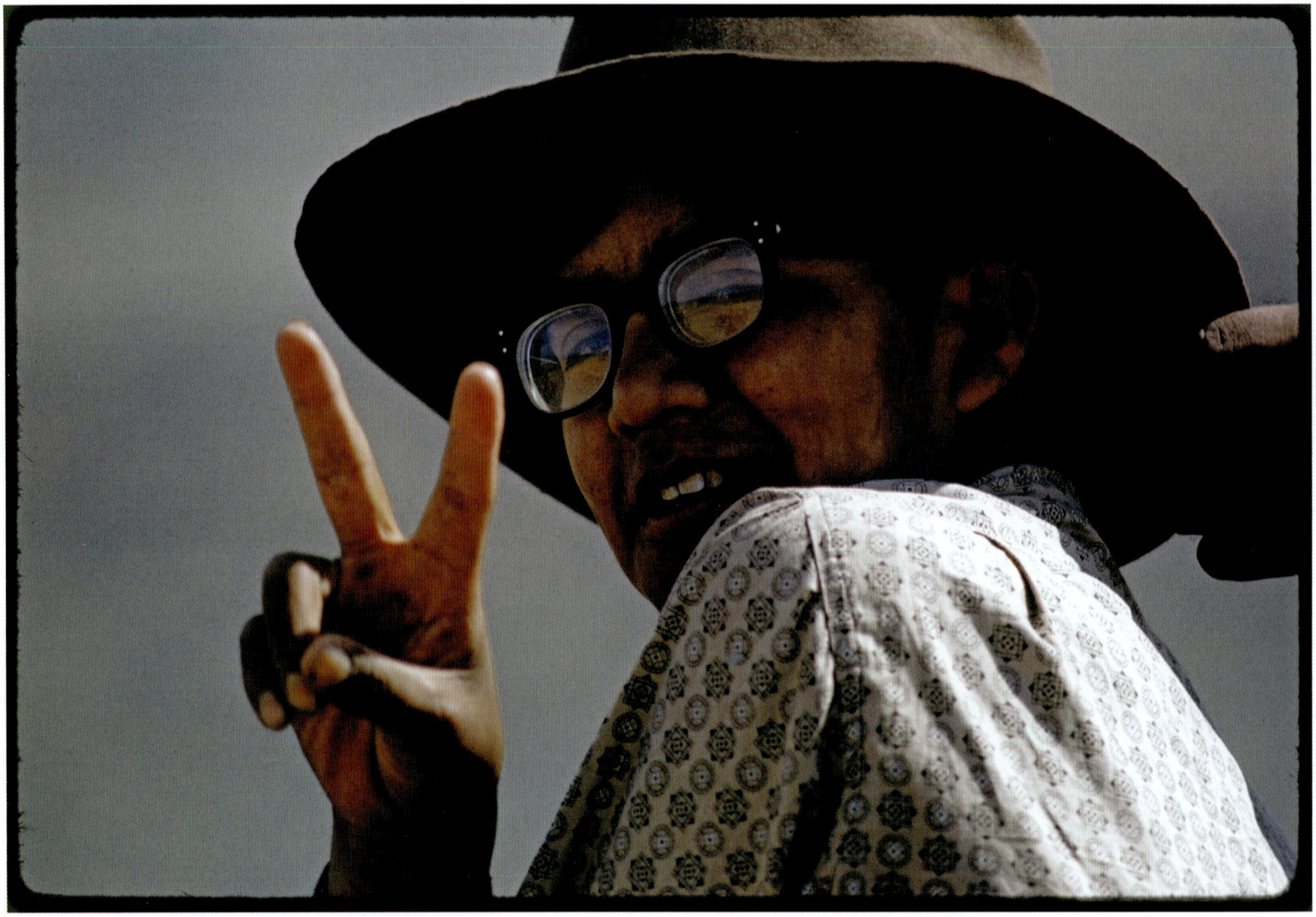

"Spring roundup of Paiute-owned cattle begins at Sutcliffe, Pyramid Lake Indian Reservation. Coralling [*sic*] and branding is done in five stages around Pyramid Lake. Posing for the Photographer."

Jonas Dovydenas, Pyramid Lake Indian Reservation, Nevada, June 1973

(412-DA-10652)

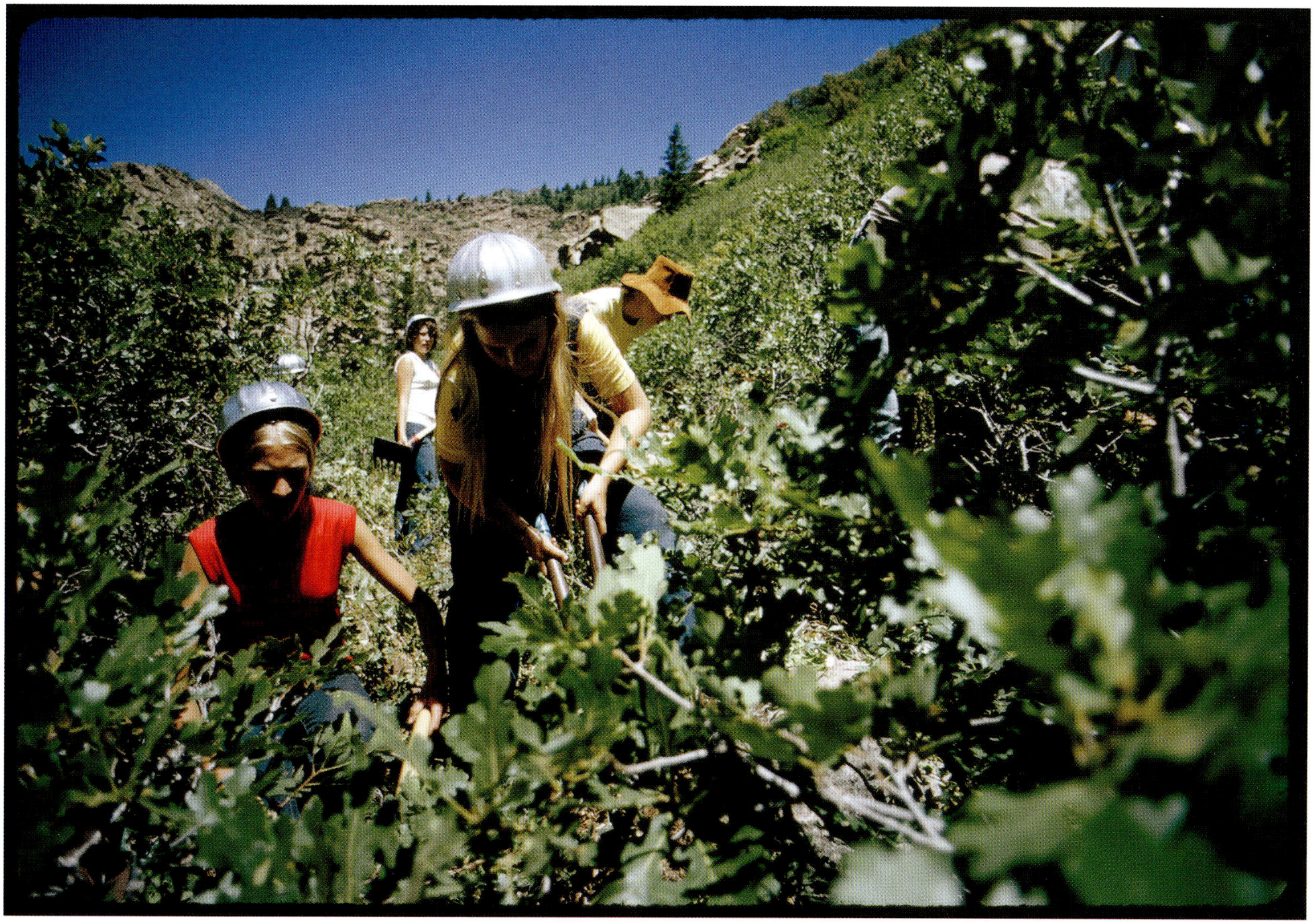

"Alta Youth Conservation Corps working on trail in Big Cottonwood Canyon."

Bruce McAllister, Alta, Utah, June 1972

(412-DA-2246)

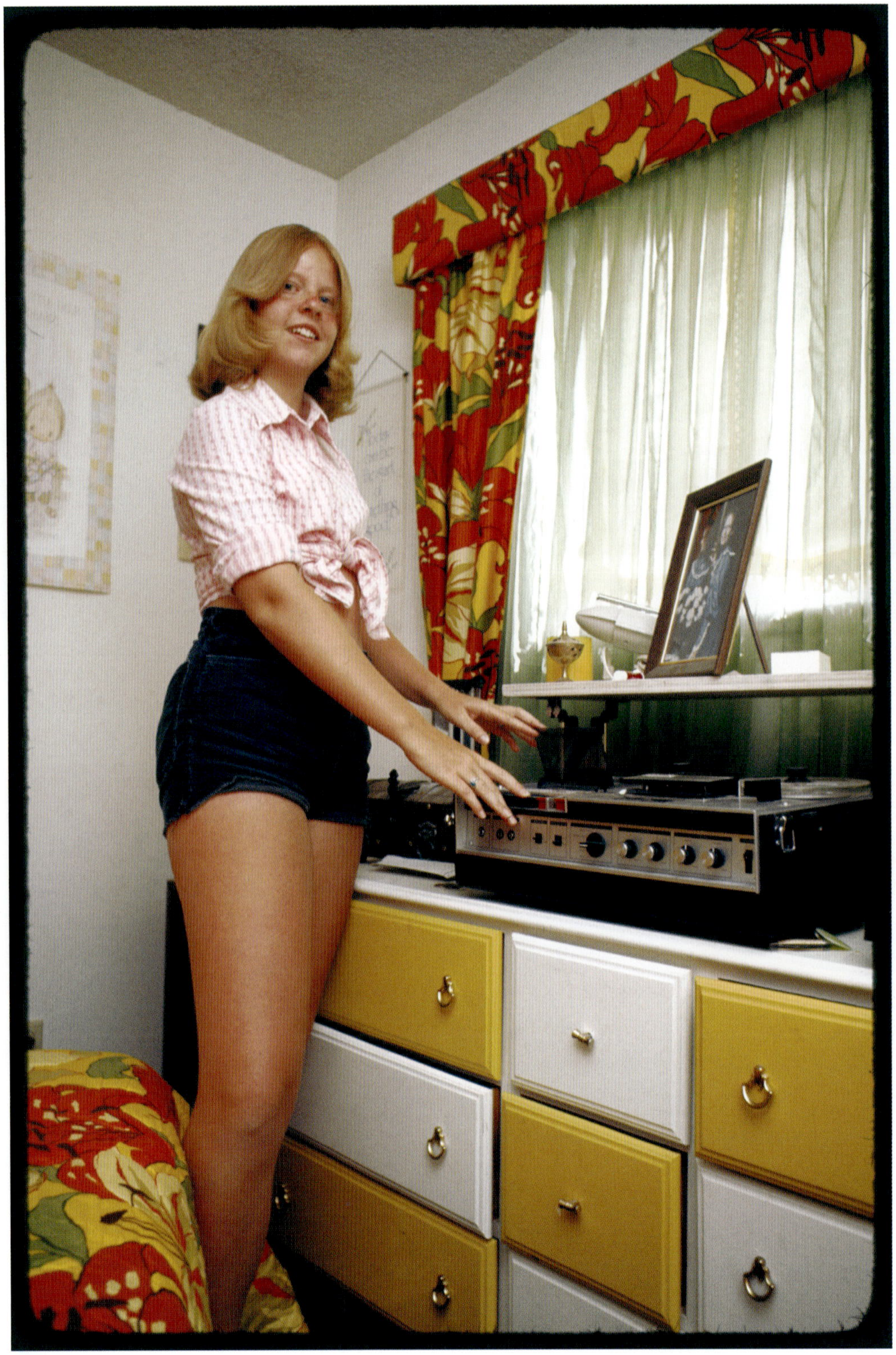

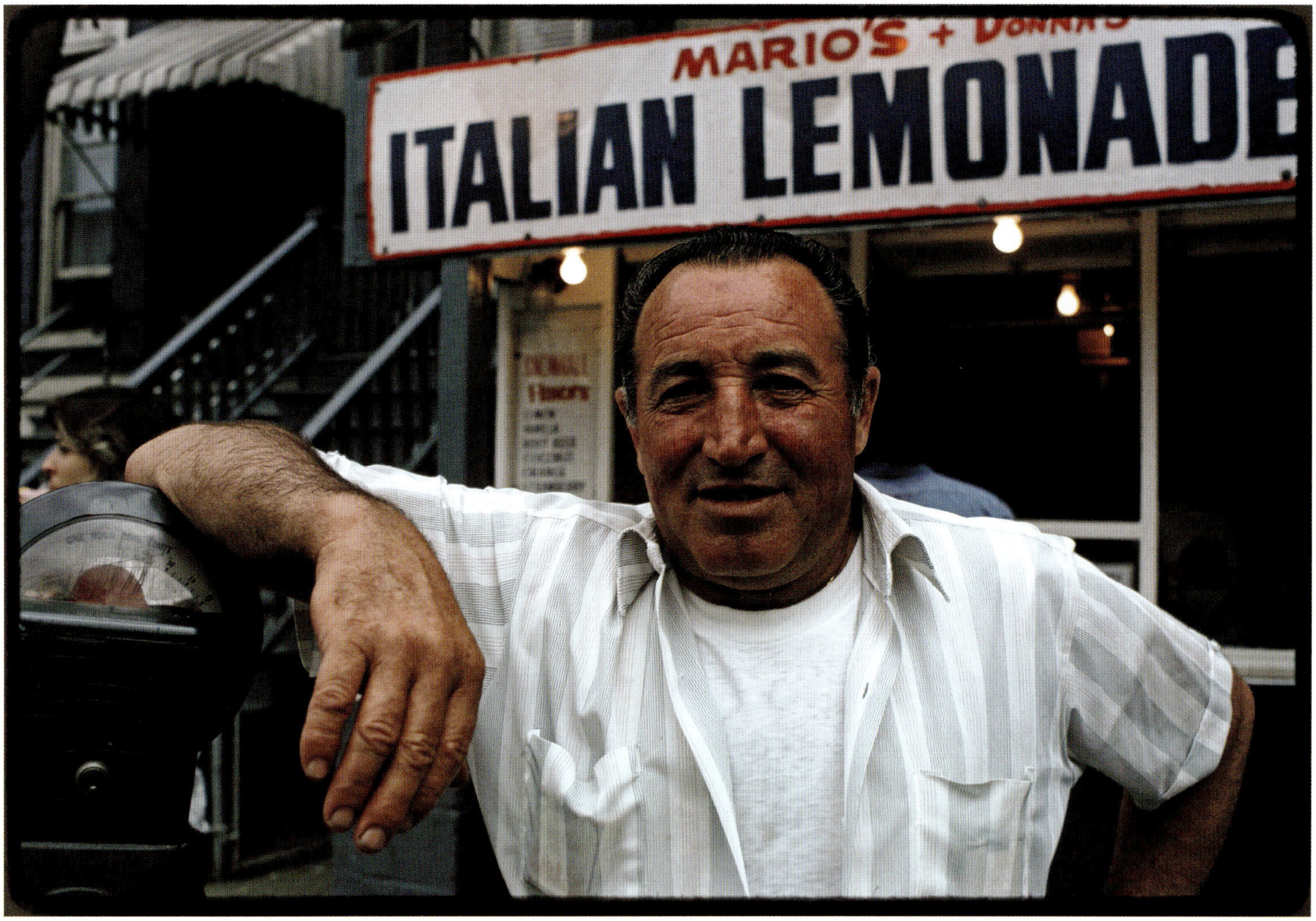

"Italian neighborhood, Chicago near southwest side."

Perry Riddle, Chicago, Illinois, July 1975

(412-DM-0153-28-13)`

Opposite: "Dana Jens in her bedroom."

Holly Wiedman, Meeker, Colorado, July 1975

(412-DM-0154-29-29)

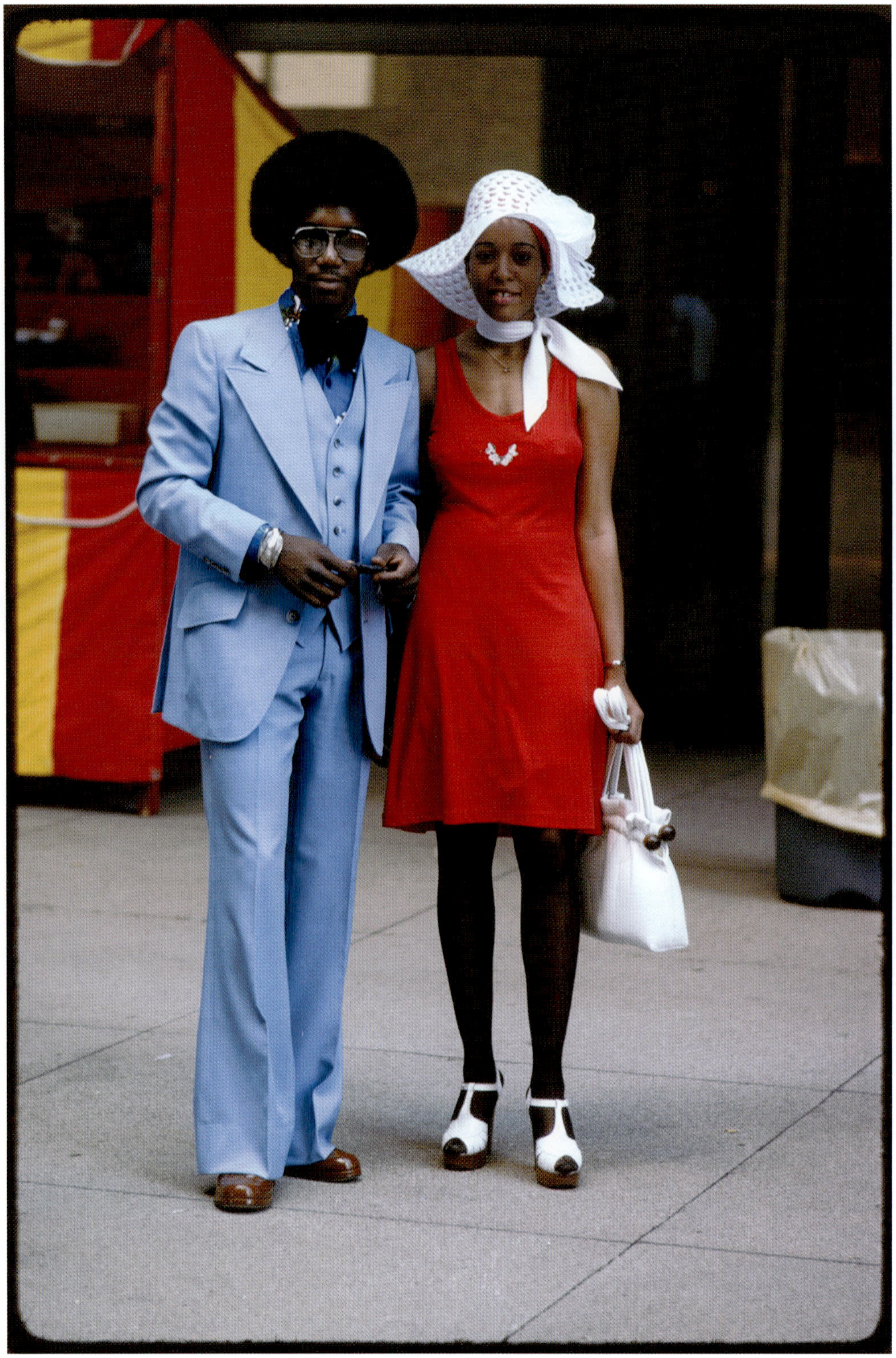

"Michigan Avenue, Chicago."

Perry Riddle, Chicago, Illinois, July 1975

(412-DM-0153-19-21)

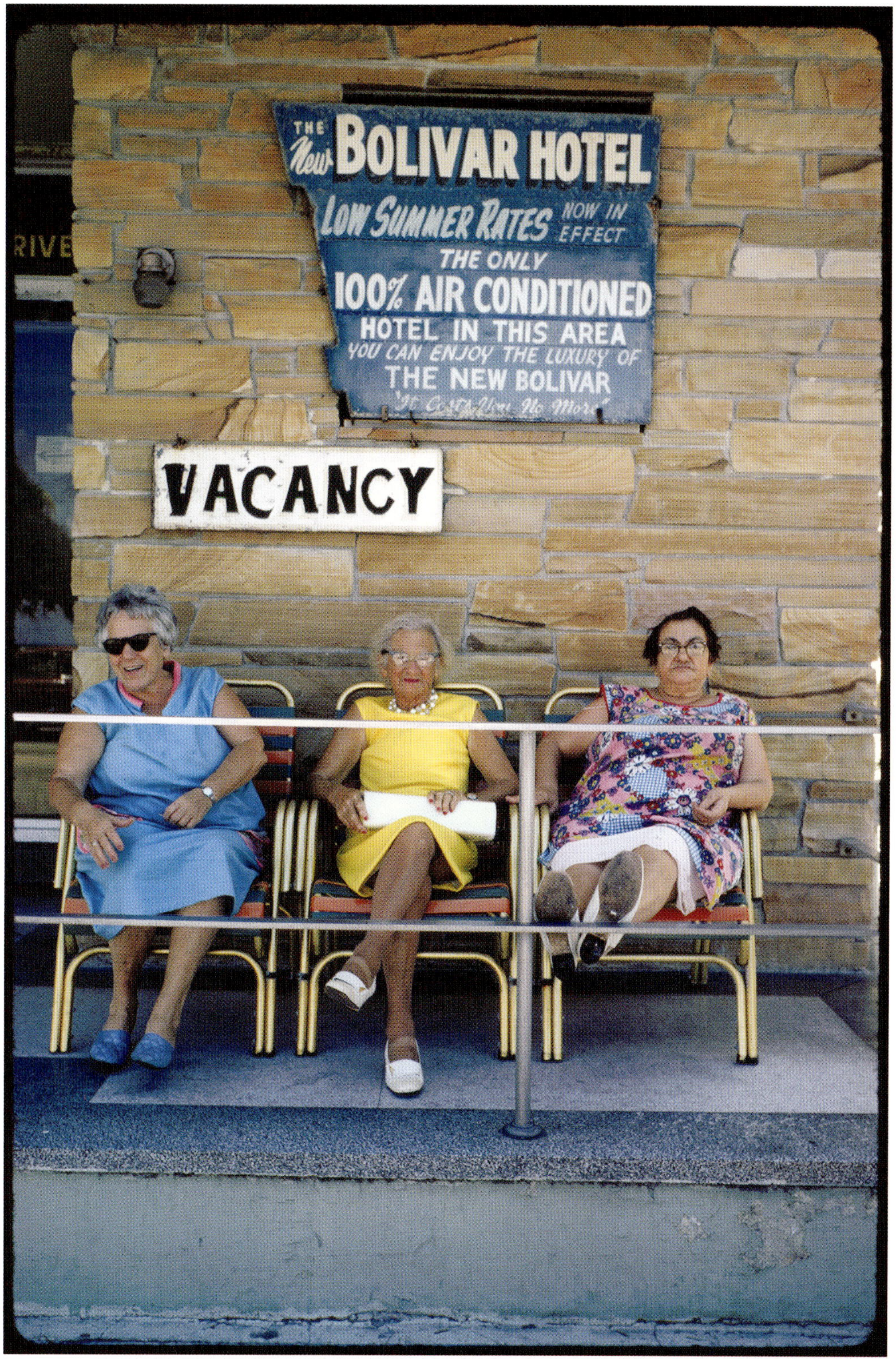

"Inexpensive retirement hotels are a hallmark of the South Beach area. A favored place is the front porch, where residents sit and chat or watch the activities on the beach."

Flip Shulke, South Beach, Miami Beach, Florida, June 1973

(412-DA-6139)

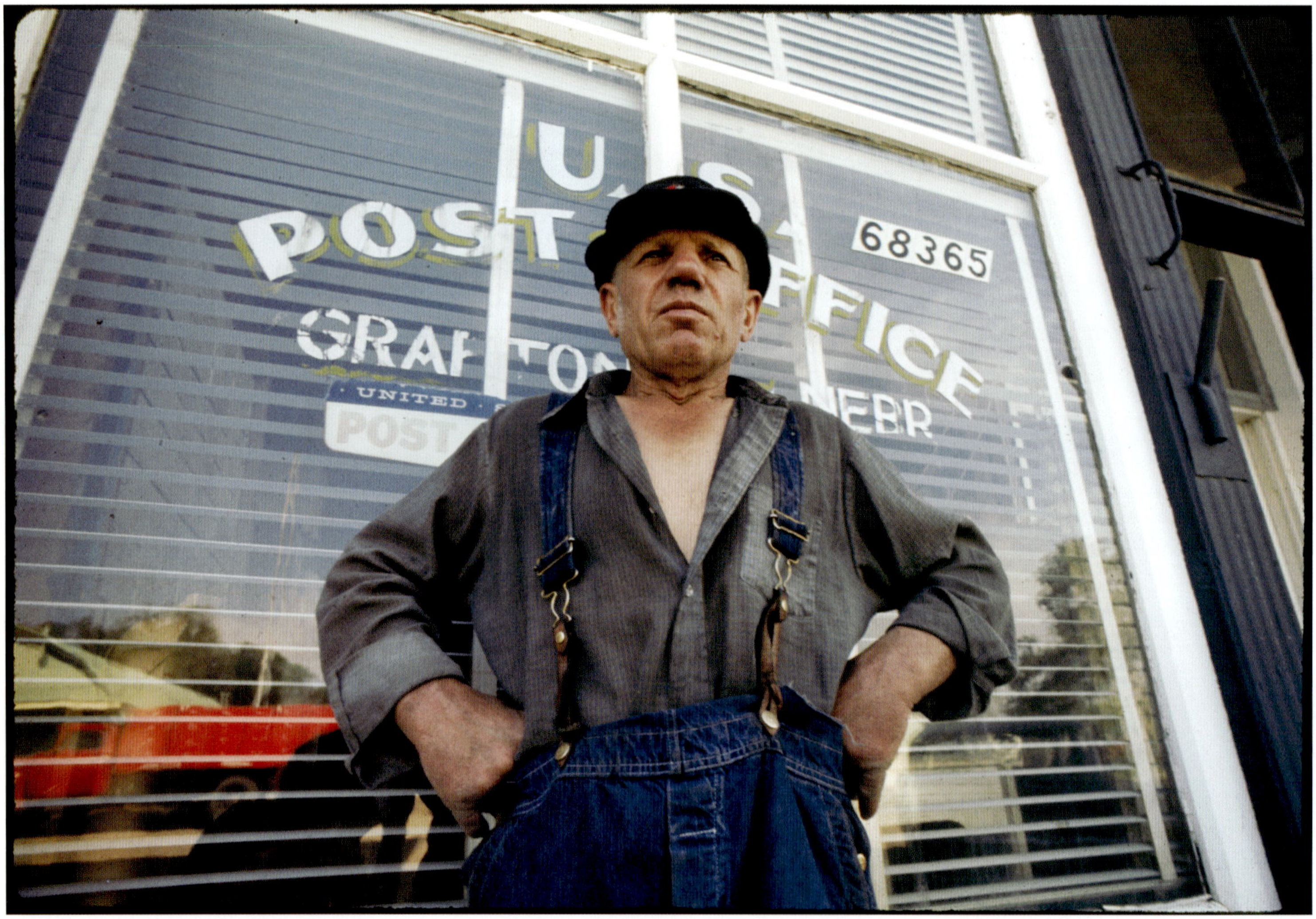

"Charlie Gross outside Grafton Post Office. A former farmer truck driver, boxer, dog raiser and salesman, he has lived in Nebraska more than half his life."

Charles O'Rear, Grafton, Nebraska, May 1973

(412-DA-4838)

Opposite: "Commuter waits for the morning Southeastern Pennsylvania Transportation Authority (SEPTA) train into downtown Philadelphia from the northwest suburbs. SEPTA has 157 bus and train lines and 2,790 route miles in the metropolitan area. State and local governments provide half the $180 million in annual operating budget, the remainder comes from fares. Some 30 million riders in 1974, out of a total of 293 million, were commuters."

Jim Pickerell, Philadelphia, Pennsylvania, May 1974

(412-DA-14313)

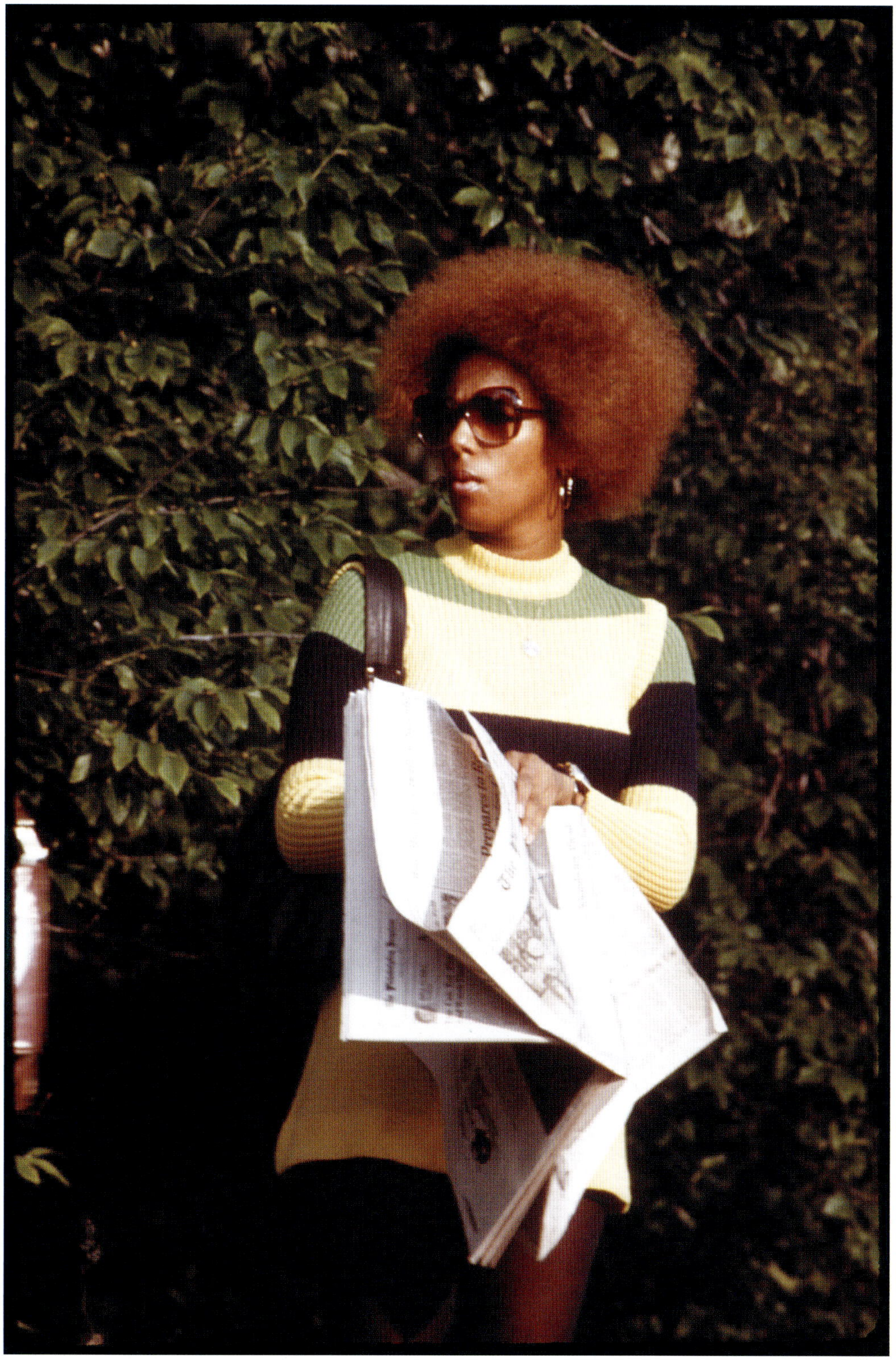

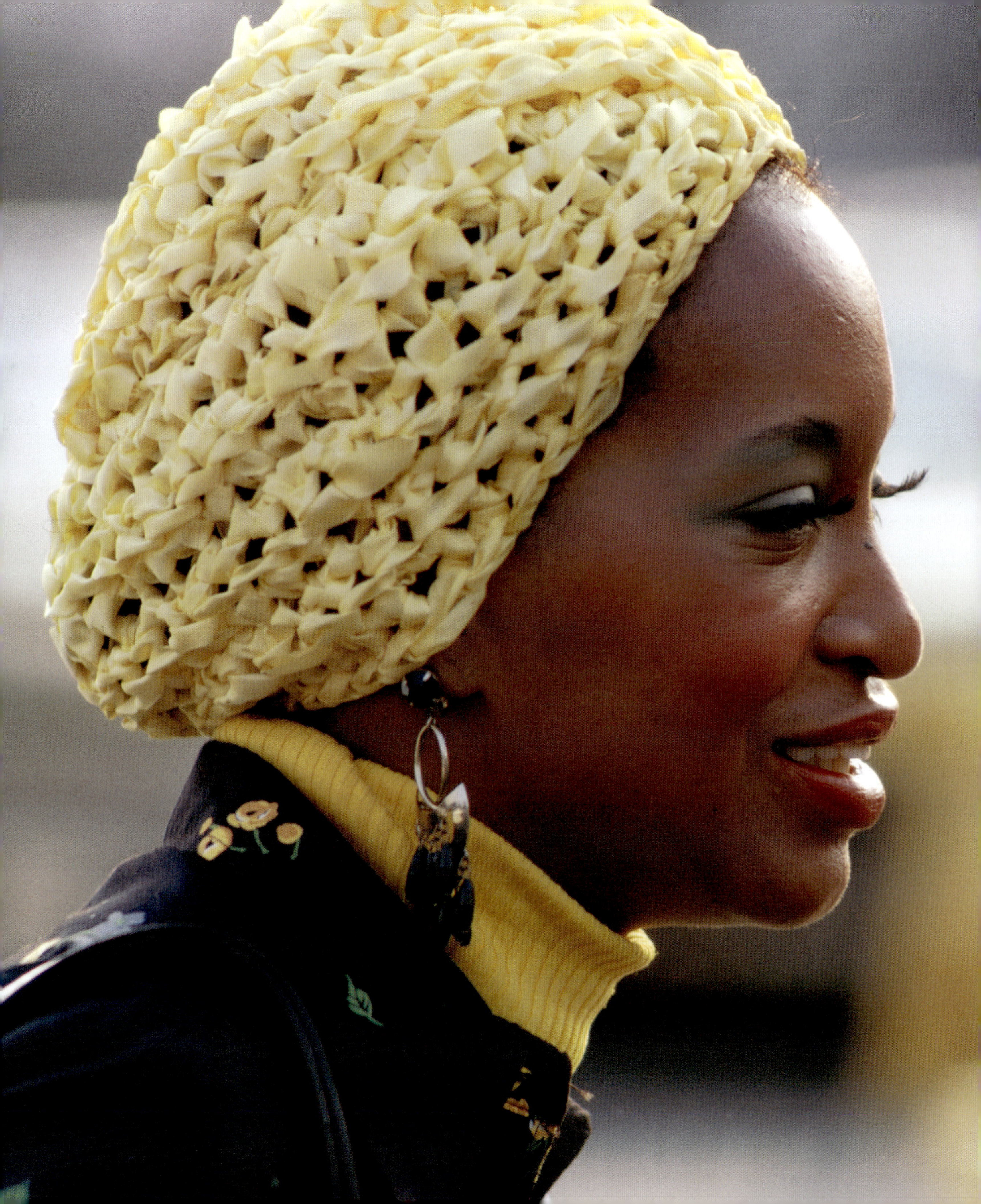

On Assignment: John H. White's Chicago

From June through October 1973 and during spring 1974, John H. White, a 28-year-old photographer with the *Chicago Daily News*, worked for DOCUMERICA photographing Chicago, especially his city's African American community. He saw his assignment as "an opportunity to capture a slice of life, to capture history." His photographs portray the difficult circumstances faced by many of Chicago's African American residents in the early 1970s, but they also catch the "spirit, love, zeal, pride, and hopes of the community." They express his belief that a "camera is a passport. It takes you into the lives of people you might otherwise never meet."

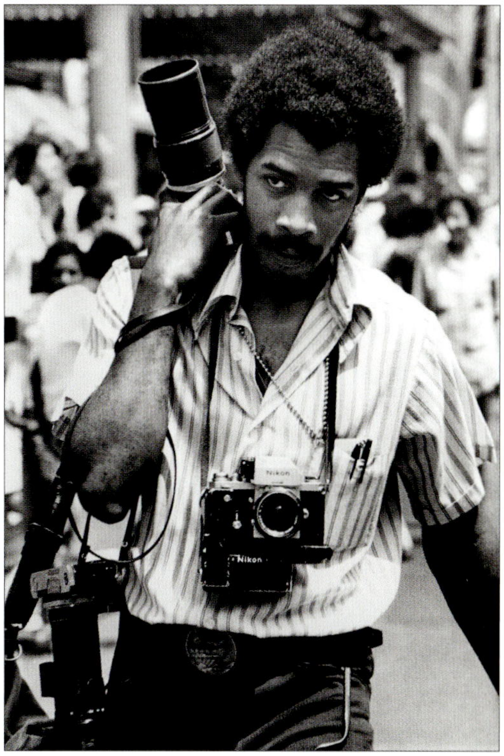

John H. White, 1977
Photograph by Jerry Tomaselli
Courtesy of John H. White

Today, White is a staff photographer with the *Chicago Sun-Times*. His assignments have included Pope John Paul II's 1979 visit to Mexico, the 1990 release of Nelson Mandela from prison, and the Rev. Jesse Jackson's peace missions to the Middle East and Yugoslavia. His publications and exhibits include two photo–essays on Archbishop Joseph Cardinal Bernardin, the *Songs of My People* book and exhibit, an exhibit at the Newseum in Washington, DC, and "Keep In Flight," a 2010–11 retrospective of his work at Chicago's Temple Gallery. He has taught at Northwestern University and continues to teach photojournalism at Columbia College Chicago. In addition to hundreds of awards, he received the Pulitzer Prize for Feature Photography in 1982.

Opposite: "Young woman soliciting funds for a Chicago organization in a shopping center parking lot. She is one of the 1.2 million black people who make up over one third of the population of Chicago. It is one of the many black faces in this project that portray life in all its seasons. The photos are portraits that reflect pride, love, beauty, hope, struggle, joy, hate, frustration, discontent, worship, and faith. She is a member of her race who is proud of her heritage."

John H. White, Chicago, Illinois, August 1973
(detail, see page 79)

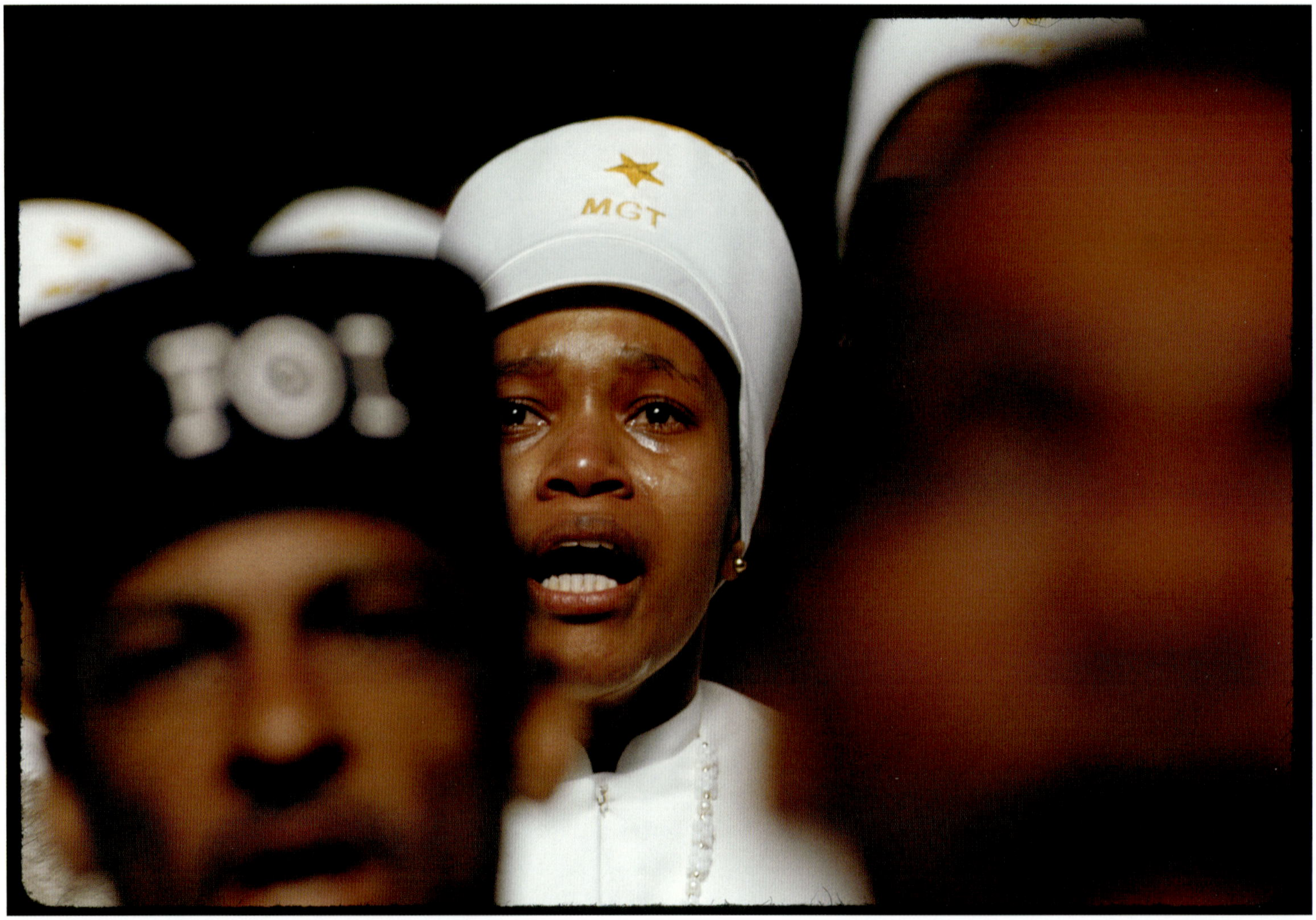

"Religious fervor is mirrored on the face of a Black Muslim woman, one of some 10,000 listening to Elijah Muhammad deliver his annual Savior's Day message in Chicago. The city is headquarters for the Black Muslims. Their $75 million Empire includes a mosque, newspaper, university, restaurants, real estate, bank, and variety of retail stores. Muhammad died February 25, 1975."

John H. White, Chicago, Illinois, March 1974
(caption written in 1975)

(412-DA-13792)

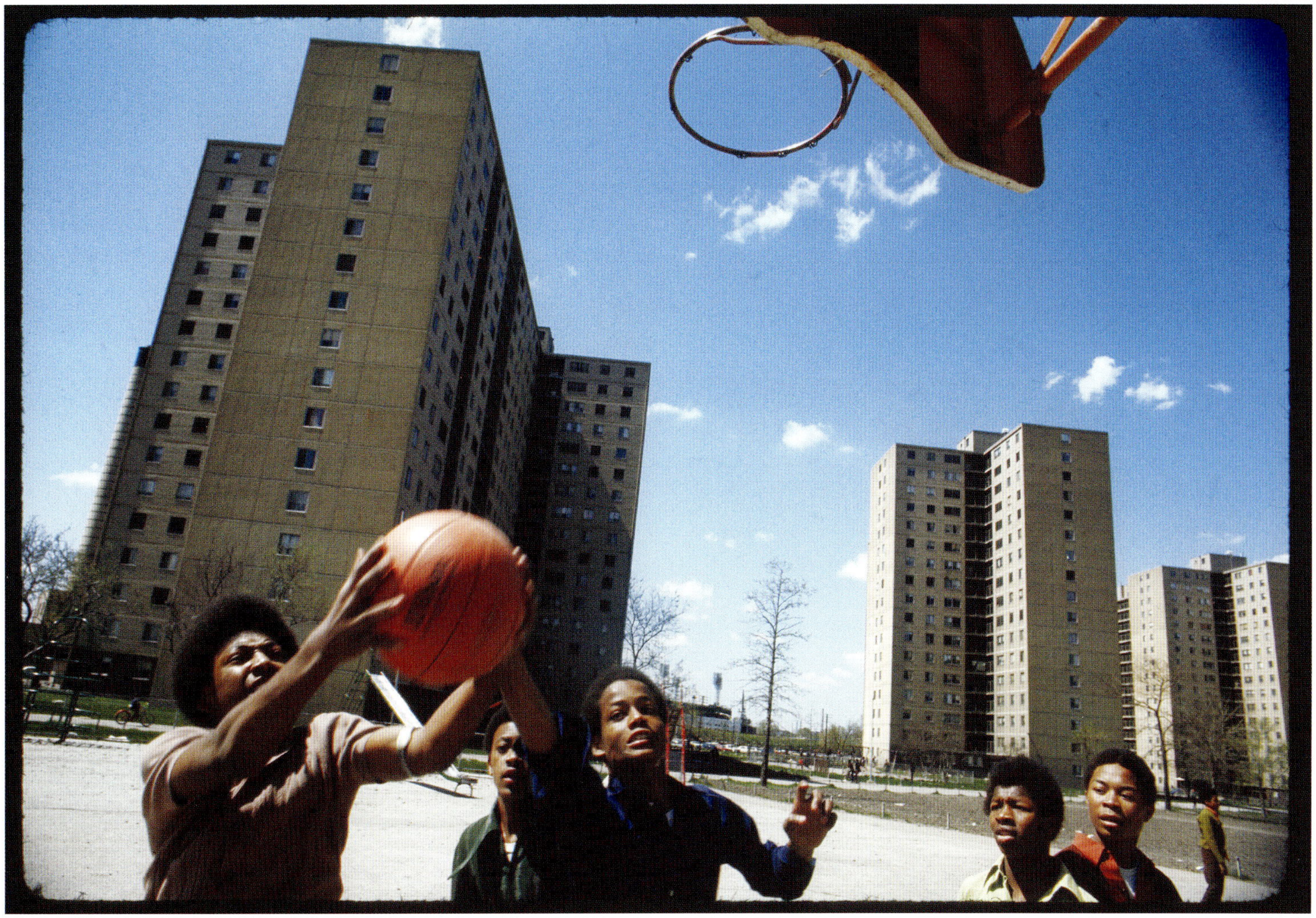

"Black youths play basketball at Stateway Gardens' highrise housing project on Chicago's South Side. The complex has eight buildings with 1,633 two and three bedroom apartments housing 6,825 persons. They were built under the U.S. Housing Acts of 1949 and 1968. They are managed by the Chicago Housing Authority which is responsible for 41,500 public housing dwellings."

John H. White, Chicago, Illinois, May 1973

(412-DA-13710)

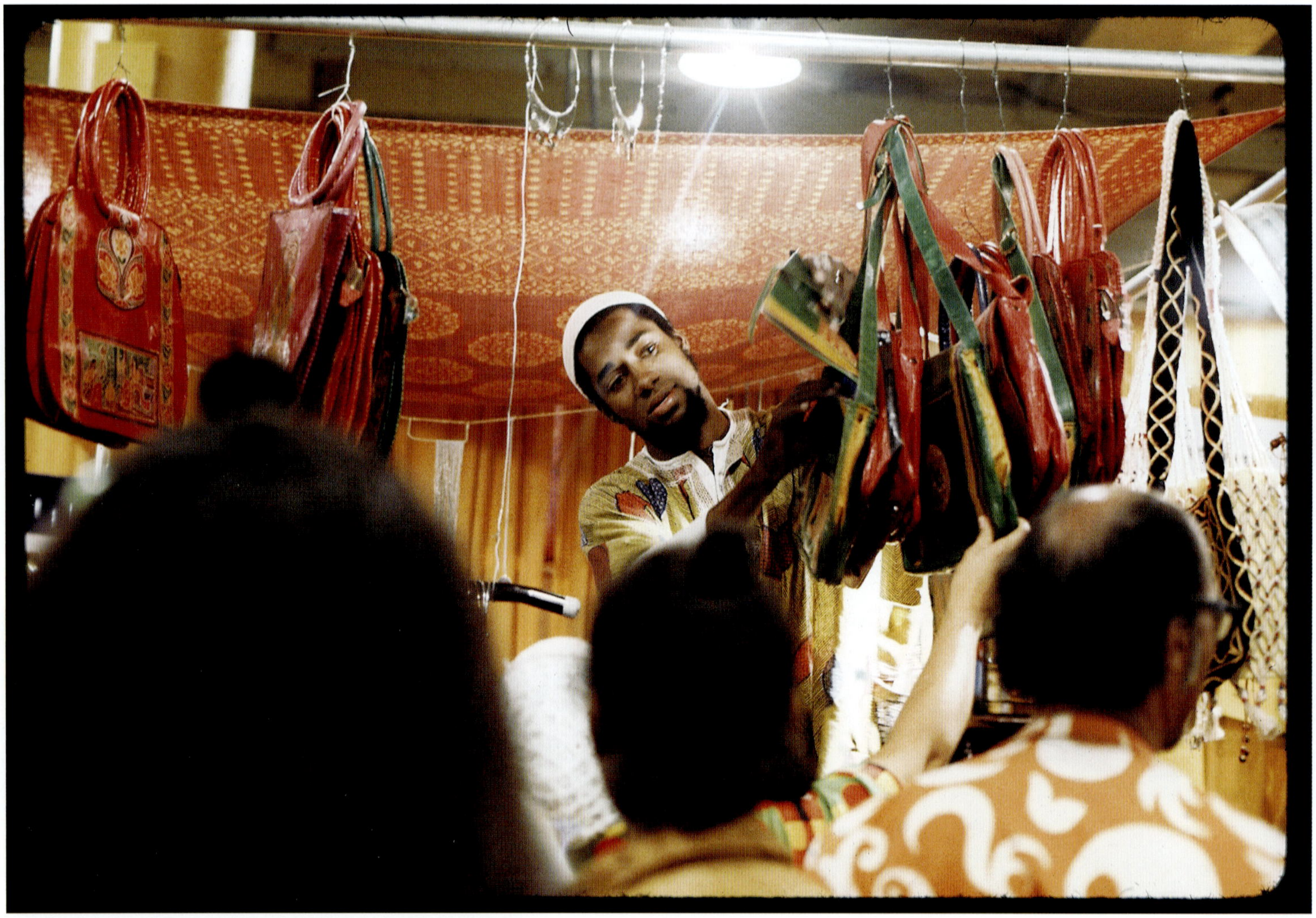

"Black products was one of the themes at the annual Black Expo held in Chicago. Also present were black education, talent, a voter registration drive and other aspects of black consciousness. The aim is to make blacks more aware of their heritage and capabilities and help them towards a better life."

John H. White, Chicago, Illinois, October 1973

(412-DA-13808)

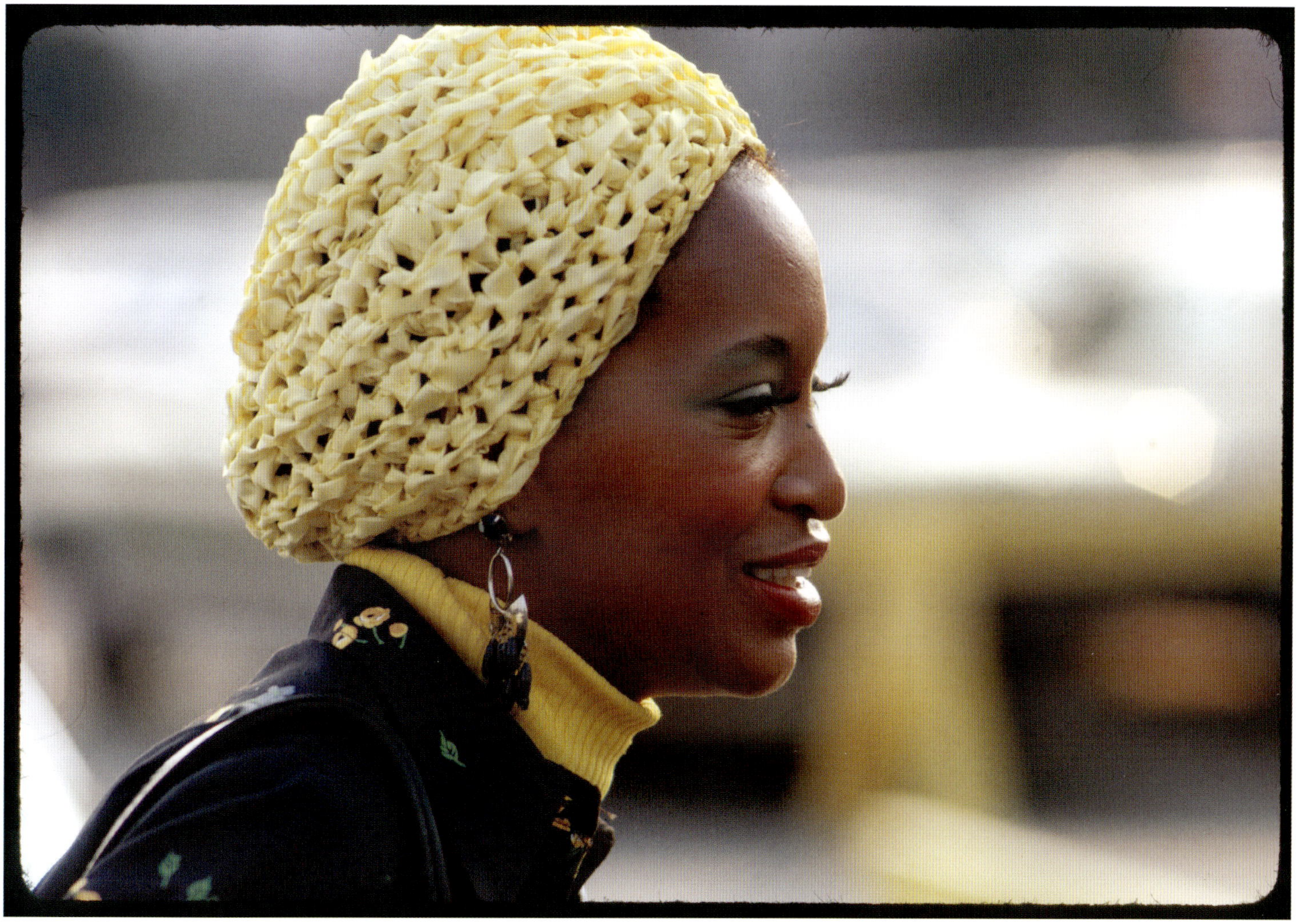

"Young woman soliciting funds for a Chicago organization in a shopping center parking lot. She is one of the 1.2 million black people who make up over one-third of the population of Chicago. It is one of the many black faces in this project that portray life in all its seasons. The photos are portraits that reflect pride, love, beauty, hope, struggle, joy, hate, frustration, discontent, worship, and faith. She is a member of her race who is proud of her heritage."

John H. White, Chicago, Illinois, August 1973

(412-DA-13682)

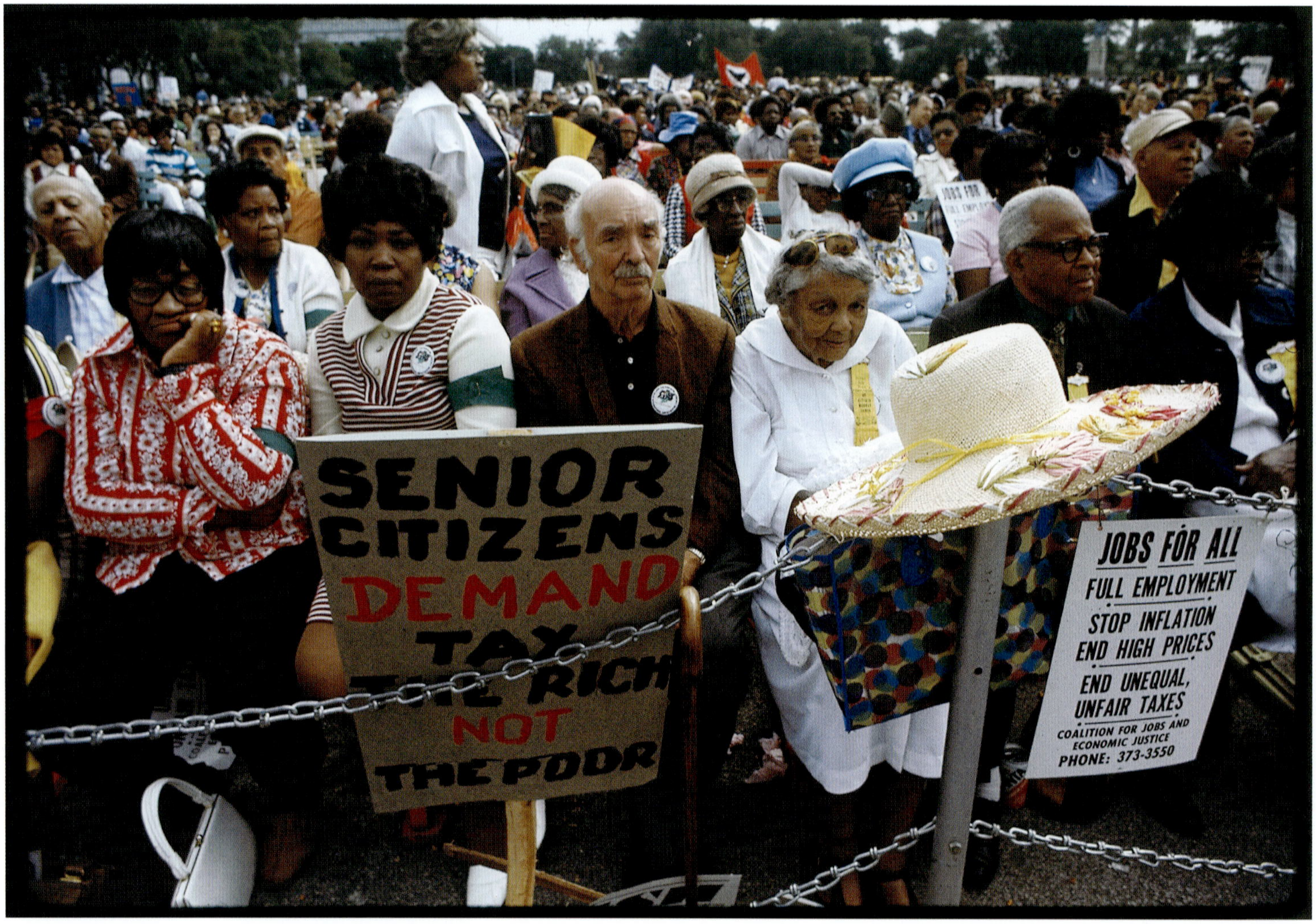

"A senior citizens march to protest inflation, unemployment and high taxes stopped along Lake Shore Drive in Chicago to hear speeches from various officials. The rally was headed by the Rev. Jesse Jackson and Operation Push."

John H. White, Chicago, Illinois, October 1973

(412-DA-13805)

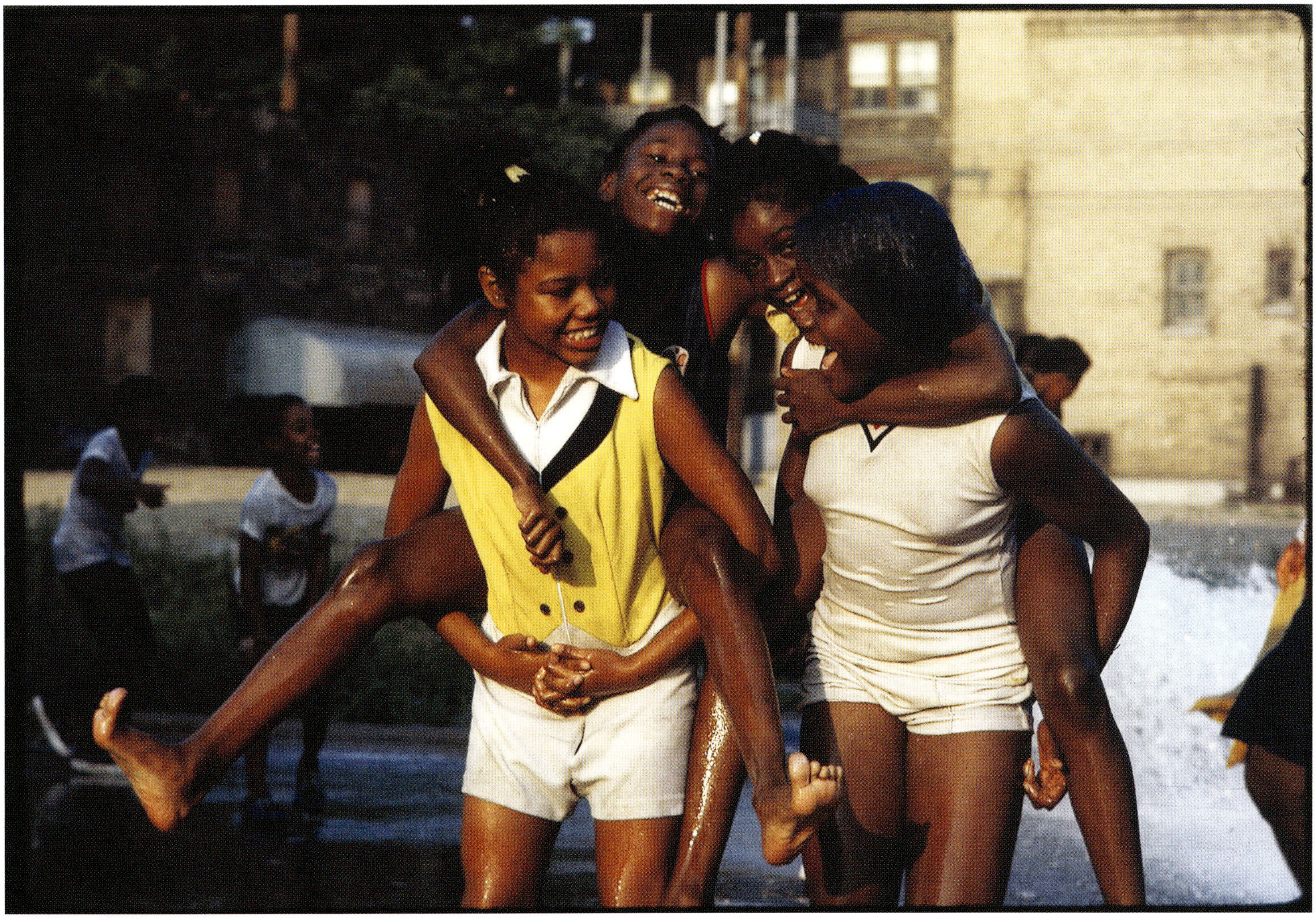

"Black youngsters cool off with fire hydrant water on Chicago's South Side in the Woodlawn community. The kids don't go to the city beaches and use the fire hydrants to cool off instead. It's a tradition in the community, comprised of very low income people. The area has high crime and fire records. From 1960 to 1970 the percentage of Chicago blacks with income of $7,000 or more jumped from 26% to 58%."

John H. White, Chicago, Illinois, June 1973

(412-DA-13684)

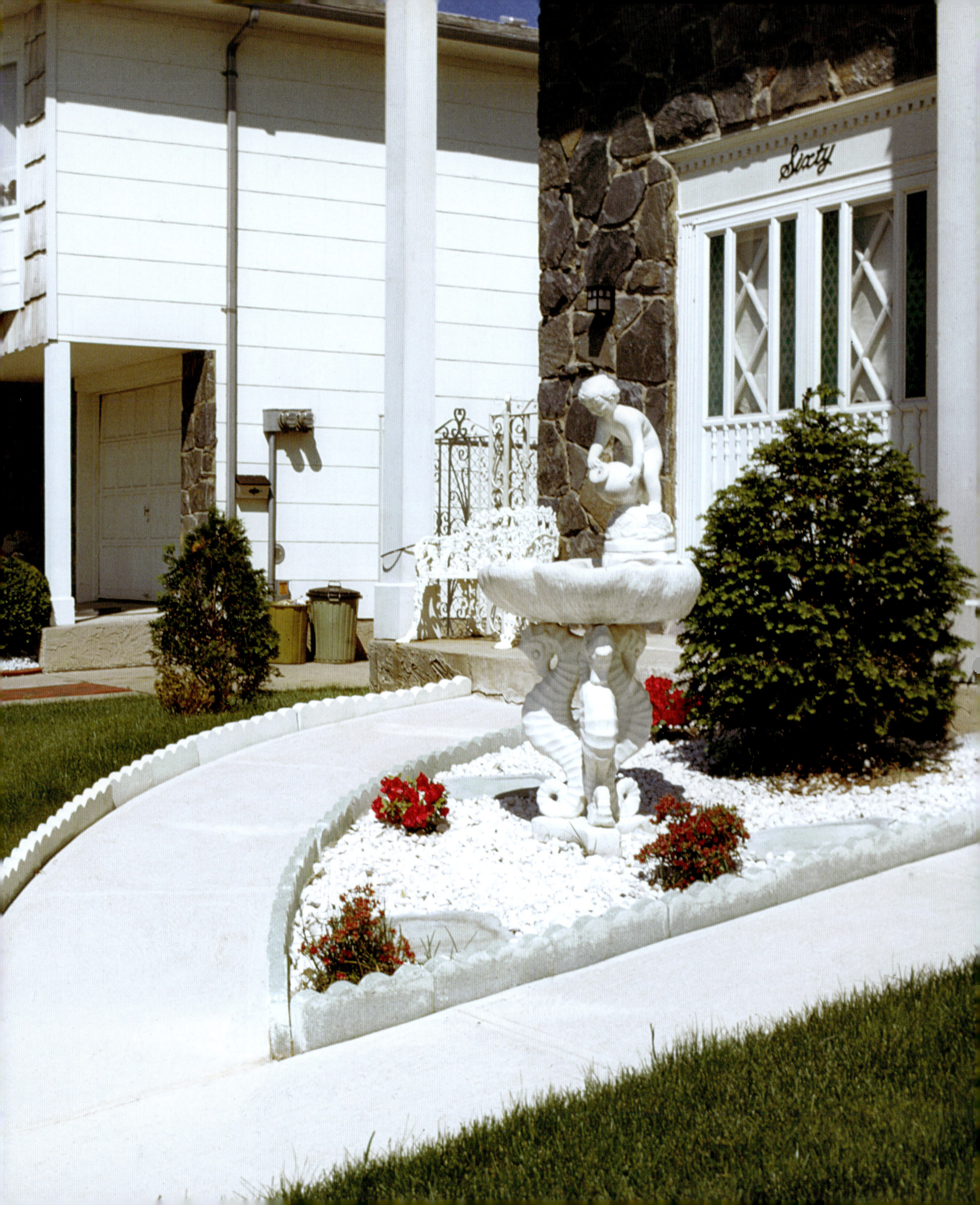
Sixty

Pave Paradise

"Don't it always seem to go that you don't
know what you've got till it's gone.
They paved paradise and put up a parking lot."
"Big Yellow Taxi," Joni Mitchell, 1970

Looking across the American landscape of the early 1970s, several features would catch your eye. Suburbs were affluent and growing, but cities were in decline—financially pressured, poverty-stricken, and crime-ridden. Places of great natural beauty continued to inspire, but they were often threatened by development and environmental damage. Americans still idealized small towns, but those communities too were challenged by pollution, economic pressures, and population loss. Urban "renewal" often resulted in the loss of culturally distinctive neighborhoods.

The automobile's influence was everywhere. Cars caused congestion, polluted the environment, consumed oil, and encouraged sprawl. They demanded networks of roads to service the nation's shopping centers, factories, schools, parks, and homes. And Americans *were* on the move: more of us were settling in the "sunbelt" and traveling further for work or leisure.

DOCUMERICA photographs depicted this fragmented landscape in its many varieties. They show scenic splendor and ugliness; urban revitalization and decay; and suburban sprawl and nostalgia for small-town America.

Opposite: "Great Kills Park, Staten Island"

Arthur Tress, Staten Island, New York, May 1973
(detail, see page 102)

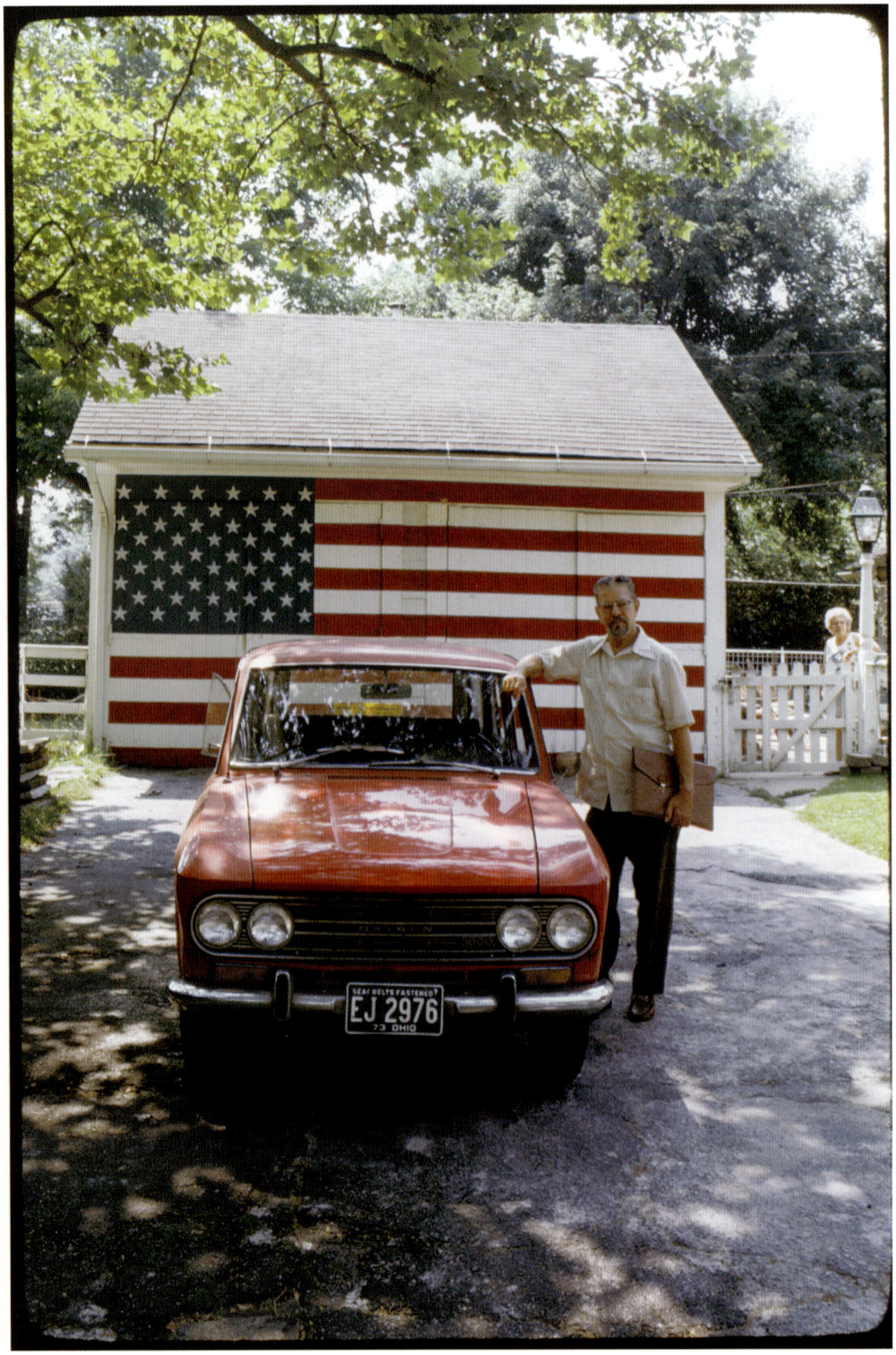
SEAT BELTS FASTENED
EJ 2976
73 OHIO

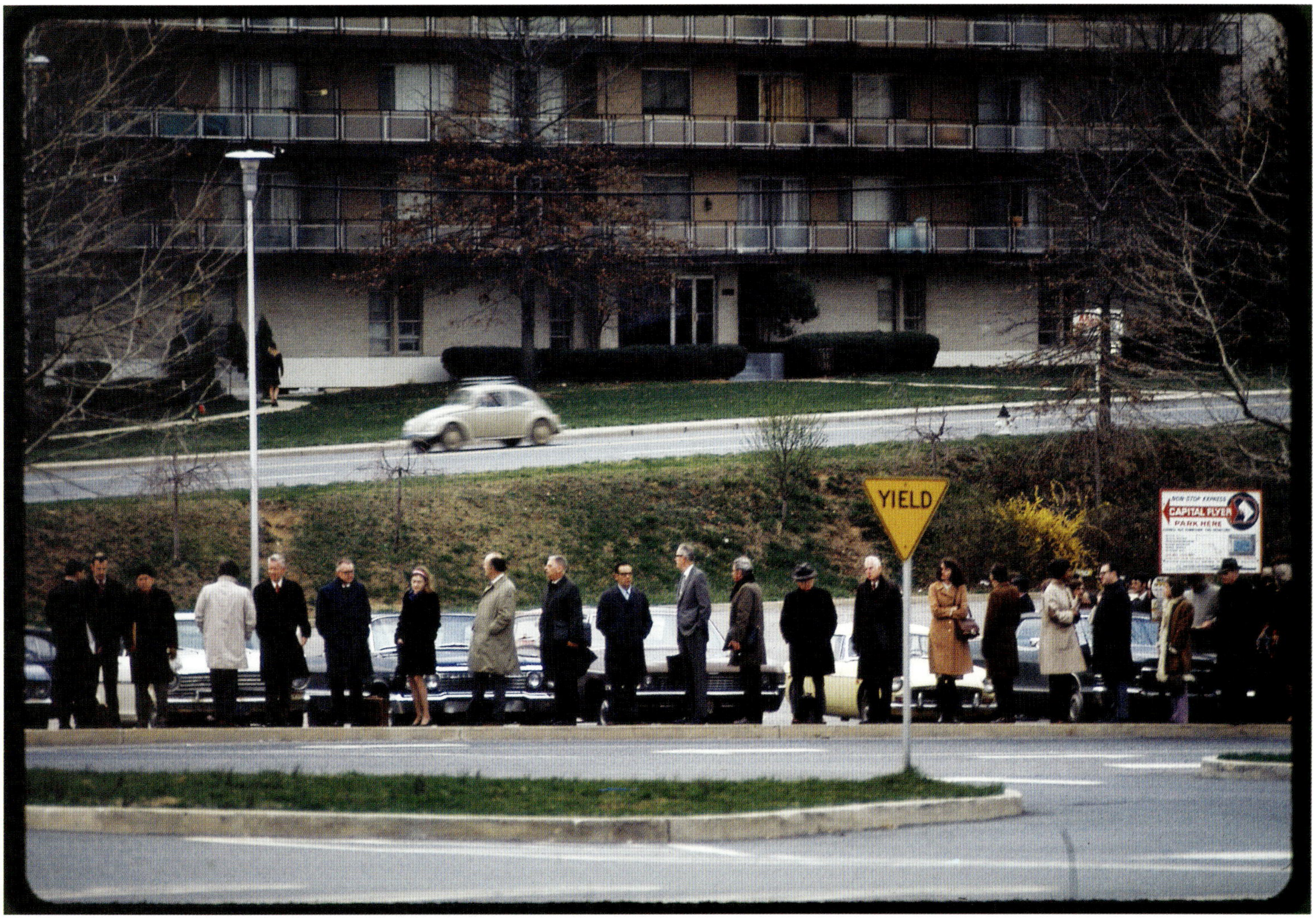

"Disciplined riders queue up for the 7:25 a.m. express bus from the Montgomery Mall shopping center (MD) into Washington, DC."

Yoichi Okamoto, Bethesda, Maryland, March 1973

(412-DA-4113)

Opposite: "The Staley's [*sic*], residents of suburban Lakewood, recently spruced up the doors of their 50-year-old garage with red, white, and blue paint."

Frank Aleksandrowicz, Lakewood, Ohio, April 1973

(412-DA-7667)

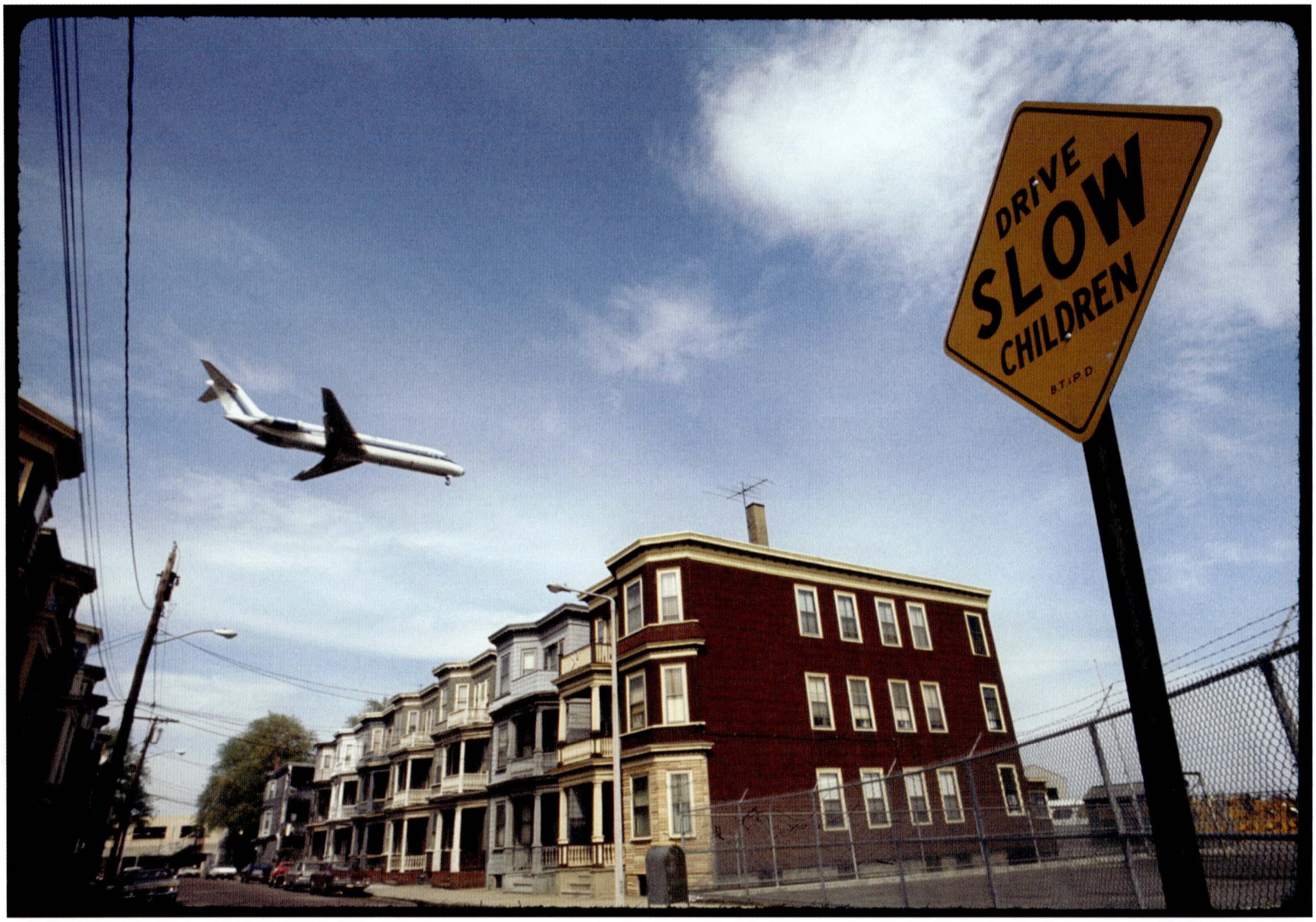

"Lovell Street homes in jet aircraft landing pattern."

Michael Philip Manheim, Boston, Massachusetts, May 1973

(412-DA-5999)

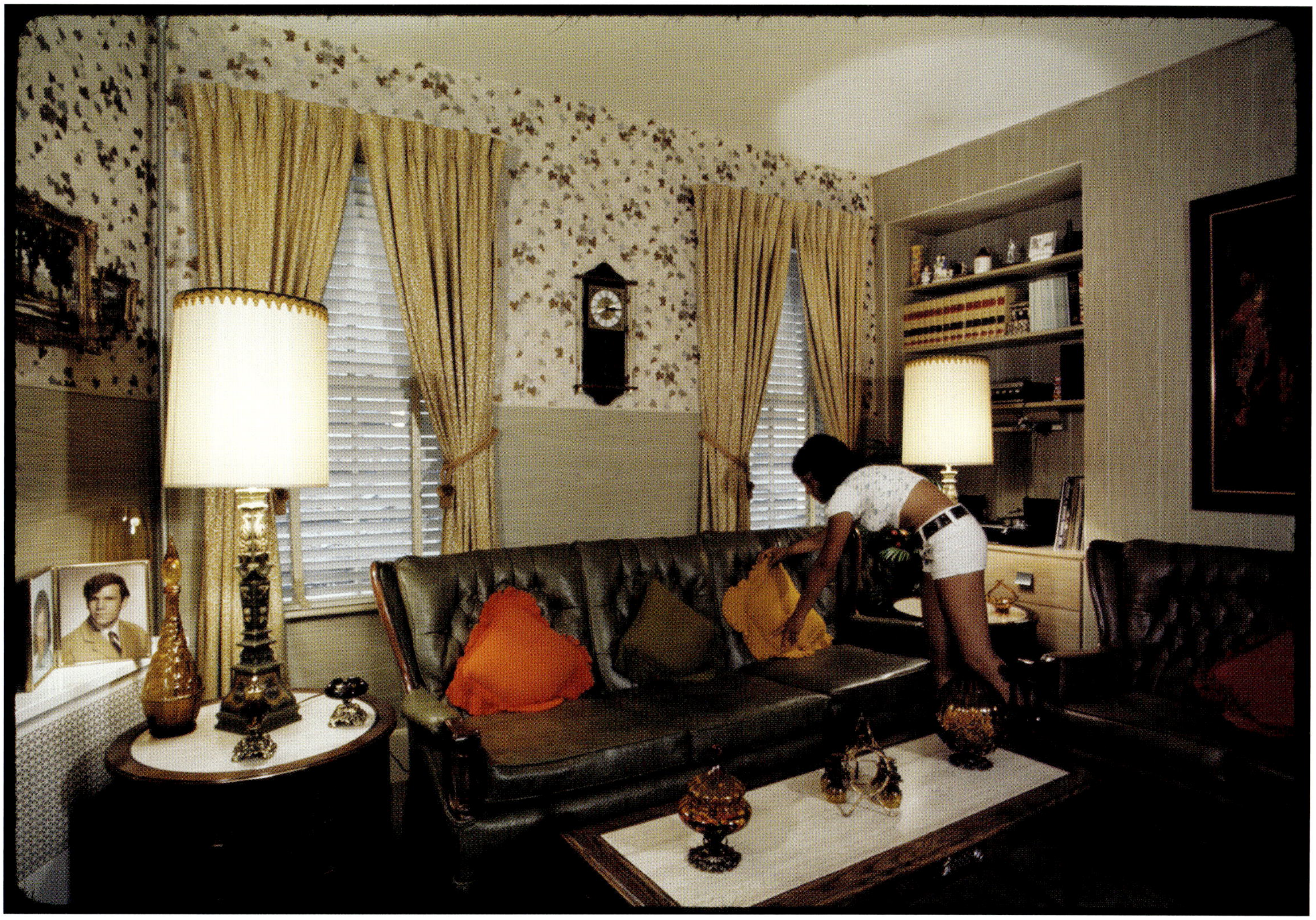

"Sandra Bruno straightens a pillow in the immaculate living room of her family's home at 39 Neptune Road."

Michael Philip Manheim, Boston, Massachusetts, July 1973

(412-DA-5953)

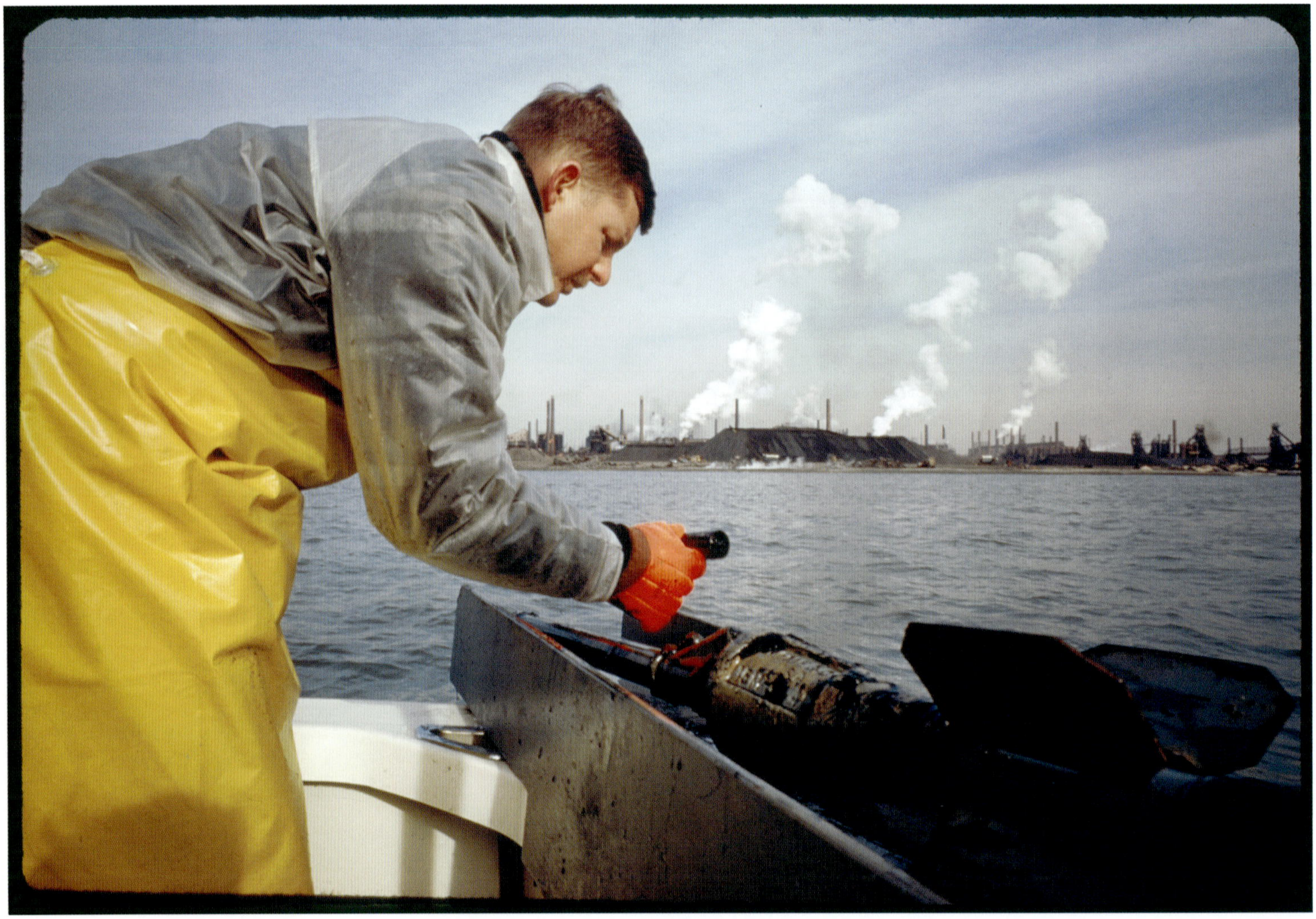

"Ron Jones, a member of the EPA Annapolis field office, takes bottom samples as part of a systematic check of Baltimore Harbor. The area being sampled here is near Bethlehem Steel's Sparrow[s] Point plant."

Jim Pickerell, Baltimore, Maryland, January 1973

(412-DA-4398)

Opposite: "Looking out at Main Street in Eastport."

Lee Lockwood, Eastport, Maine, May 1973

(412-DA-7838)

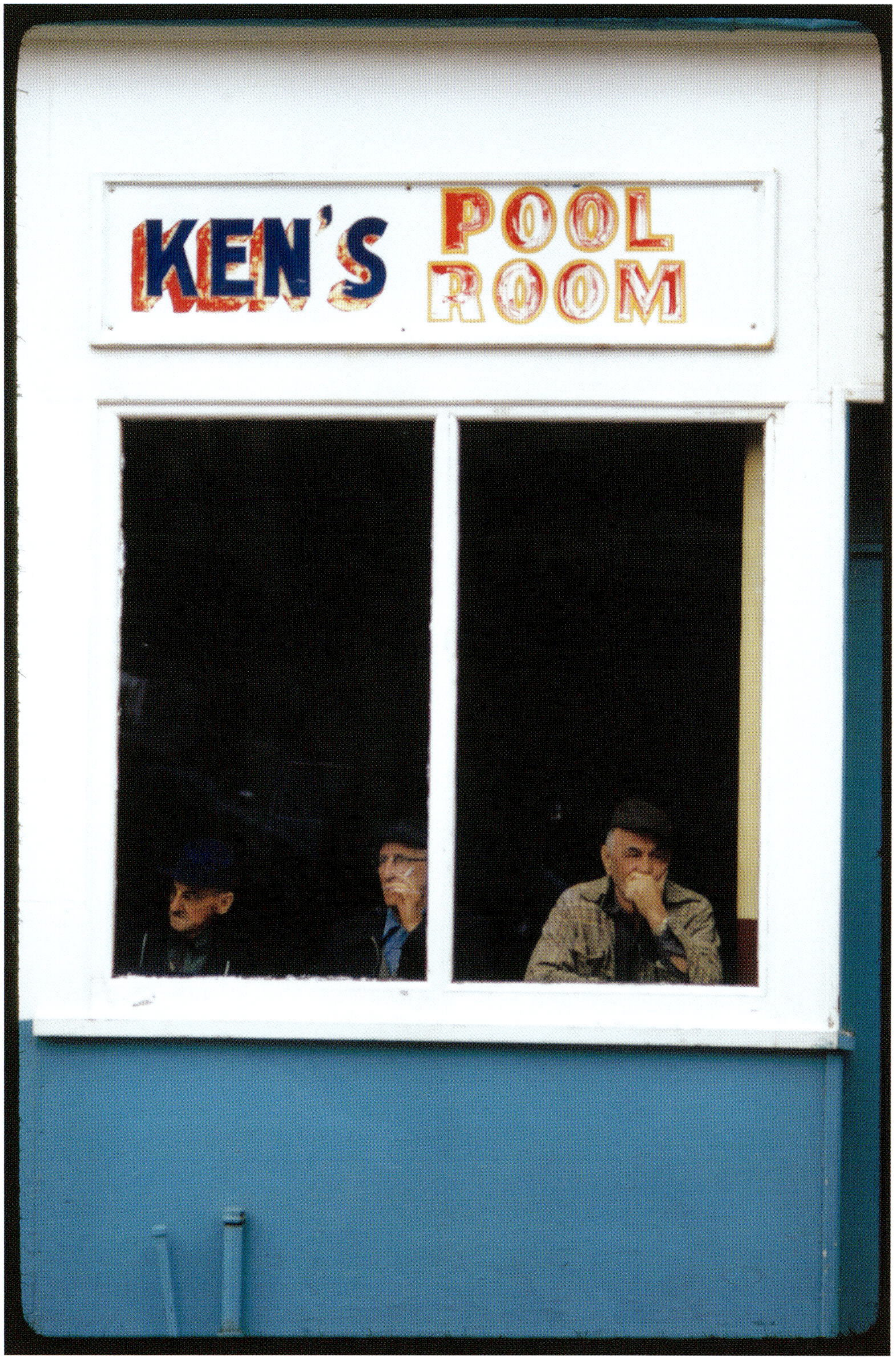
KEN'S
POOL
ROOM

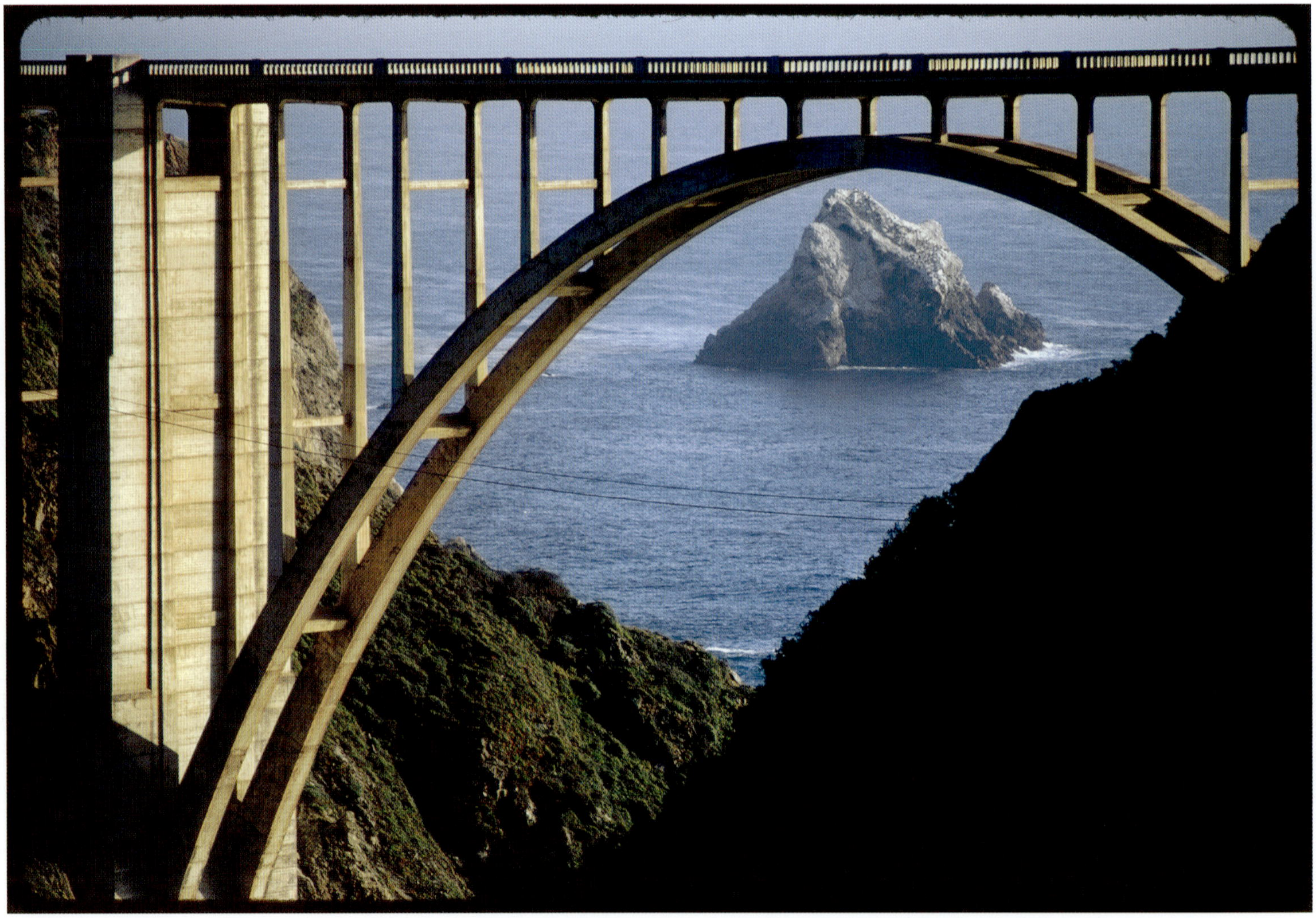

"California—Rocky Point."

Dick Rowan, California,
May 1972

(412-DA-820)

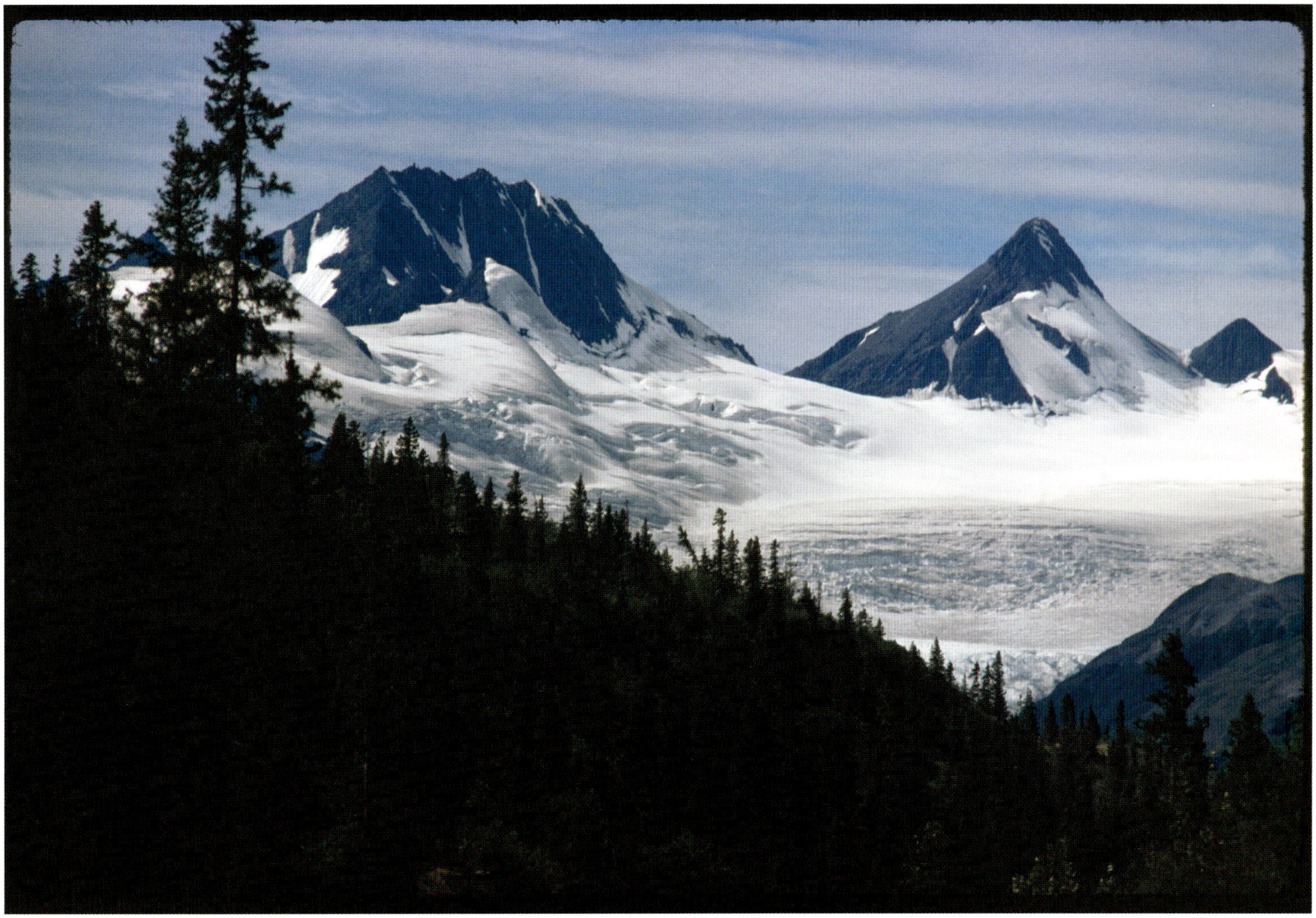

"View west parallel to pipeline alignment to Worthington Glacier which descends from 6,000-foot-high peaks almost to the Richardson Highway. The pipeline will be pushed through dense forest like seen here prior to crossing the Richardson Highway and Tsina River for the second time in seven miles. Mile 749, near the Alaska pipeline route."

Dennis Cowals, Alaska, August 1974

(412-DA-13210)

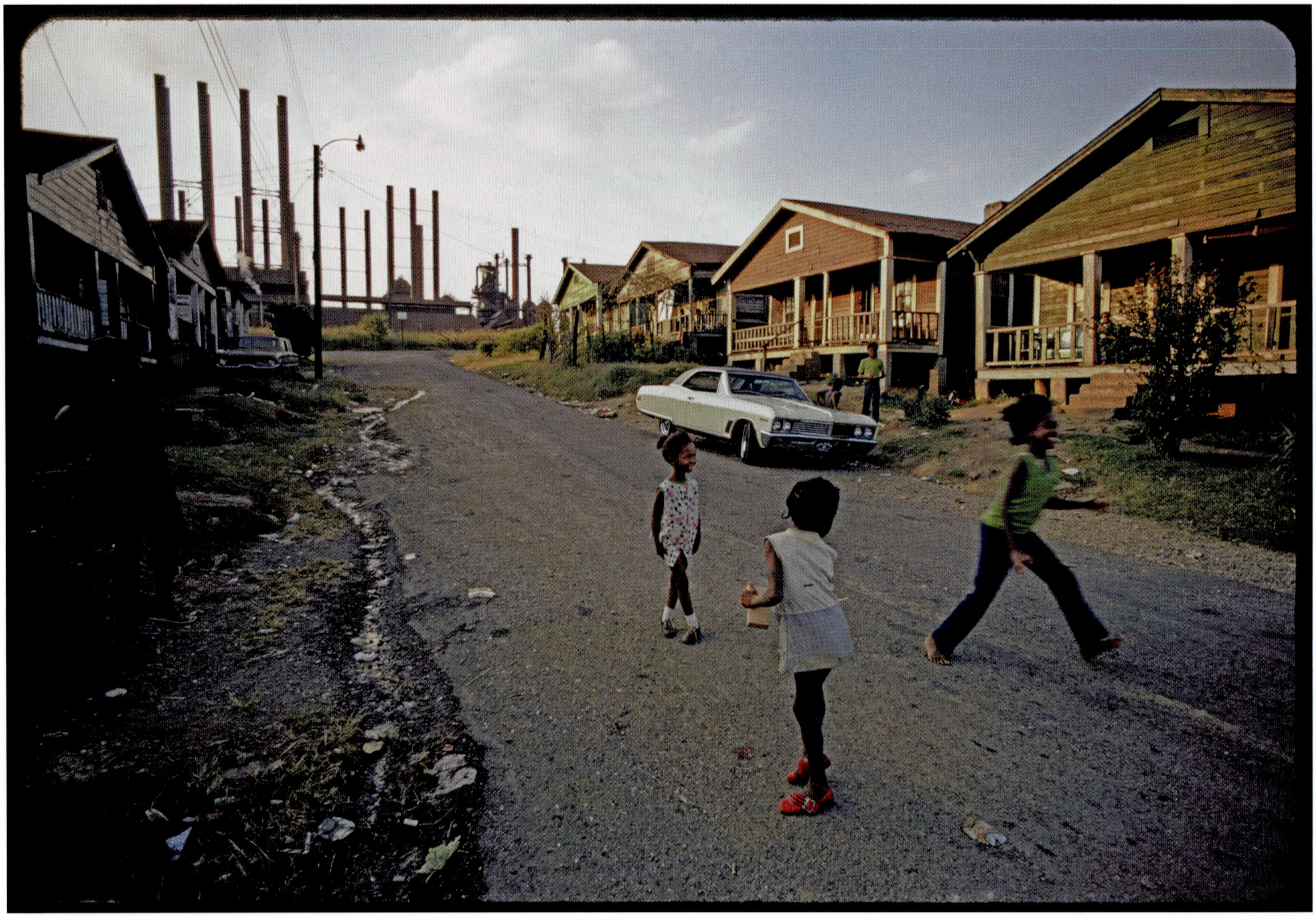

"Housing adjacent to U.S. Steel plant."

Leroy Woodson, Birmingham, Alabama, July 1972

(412-DA-2911)

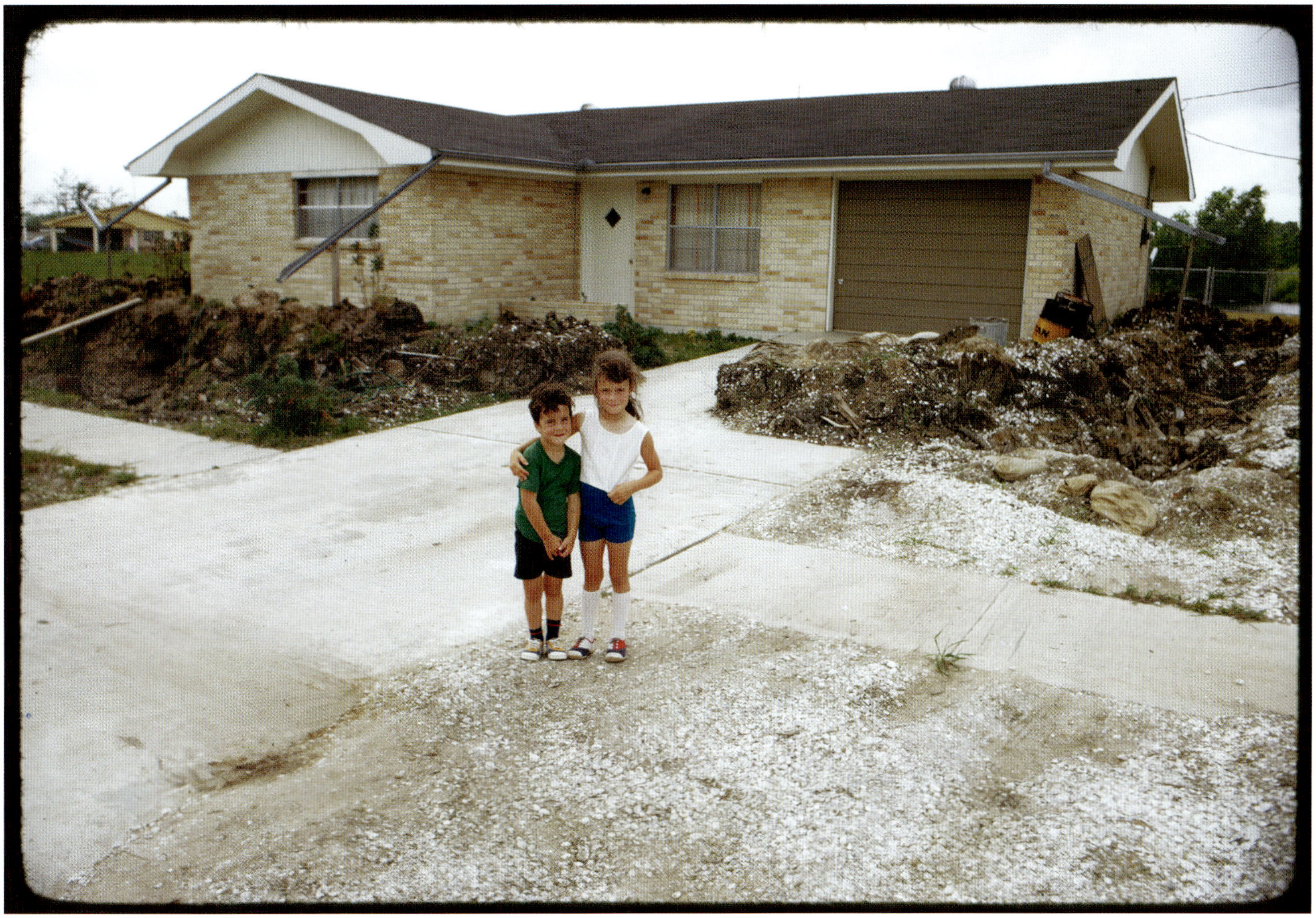

"In the spring of 1973 the Mississippi River reached it highest level in more than 150 years. Unprecedented flooding occurred throughout the river basin. Particularly affected were the marsh area below New Orleans and the entire Atchafalaya River basin. Stevensville children in front of a trenched house. Owner dug trench and formed levee to protect house from flood waters."

John Messina, Stevensville, Louisiana, May 1973

(412-DA-10345)

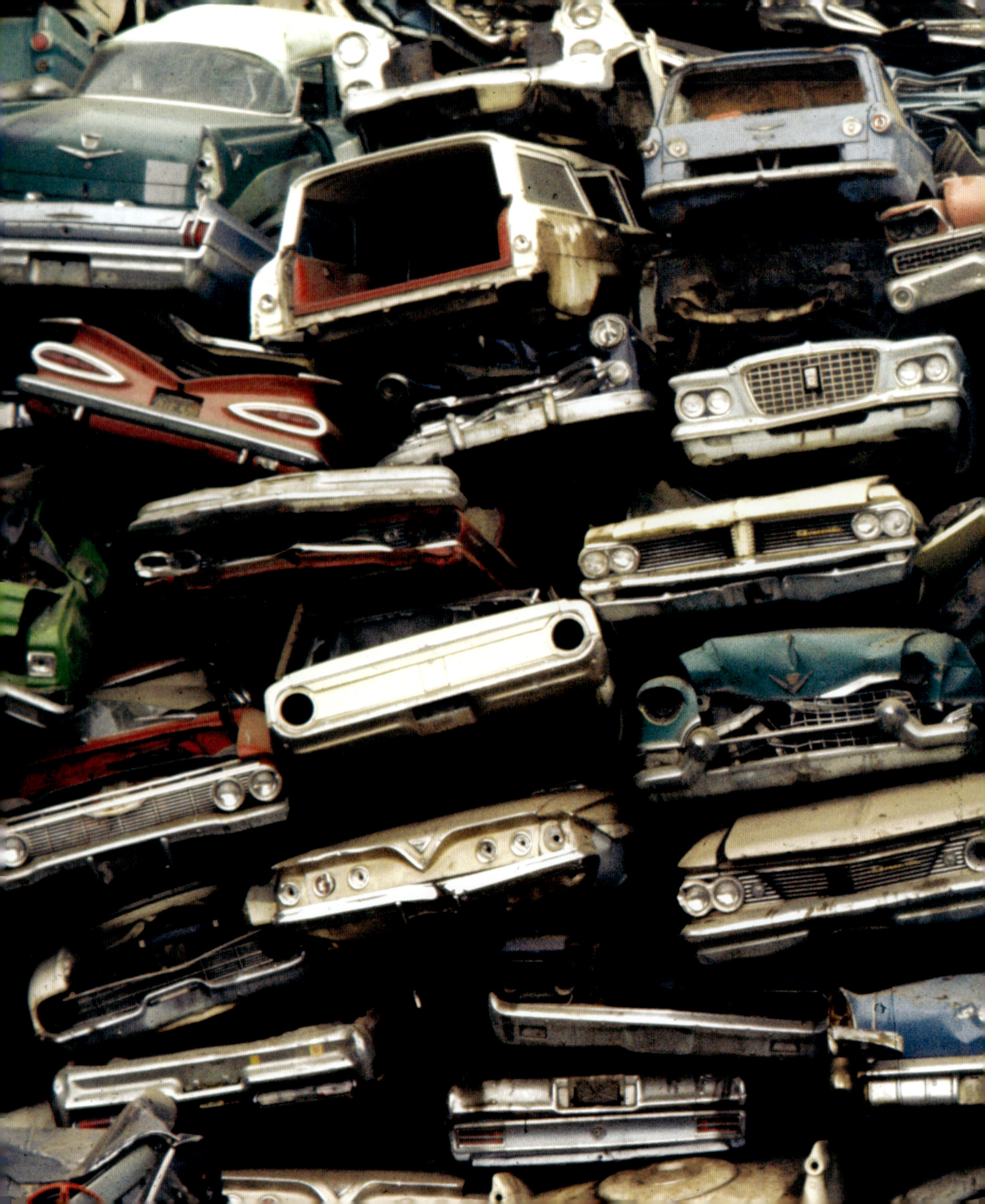

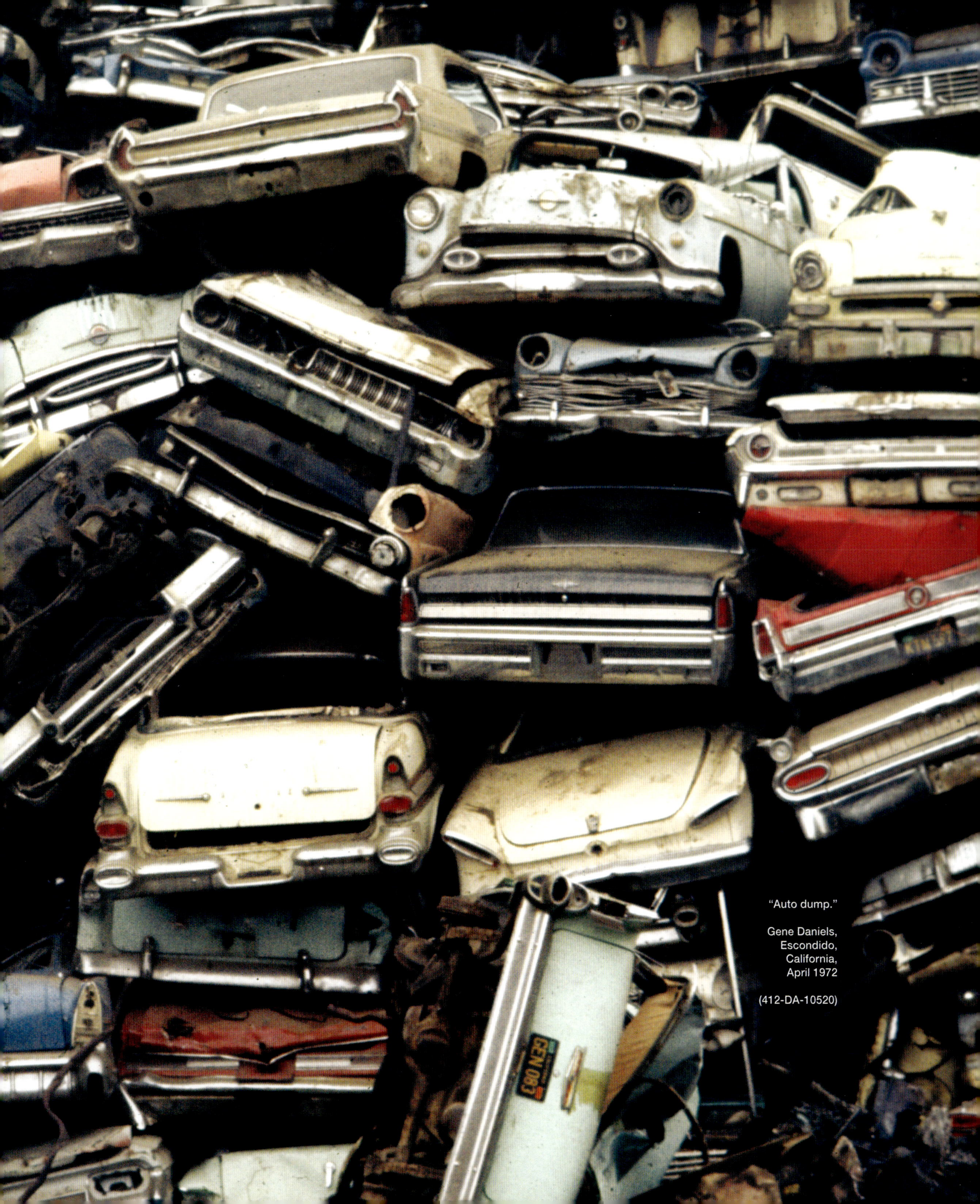

"Auto dump."

Gene Daniels,
Escondido,
California,
April 1972

(412-DA-10520)

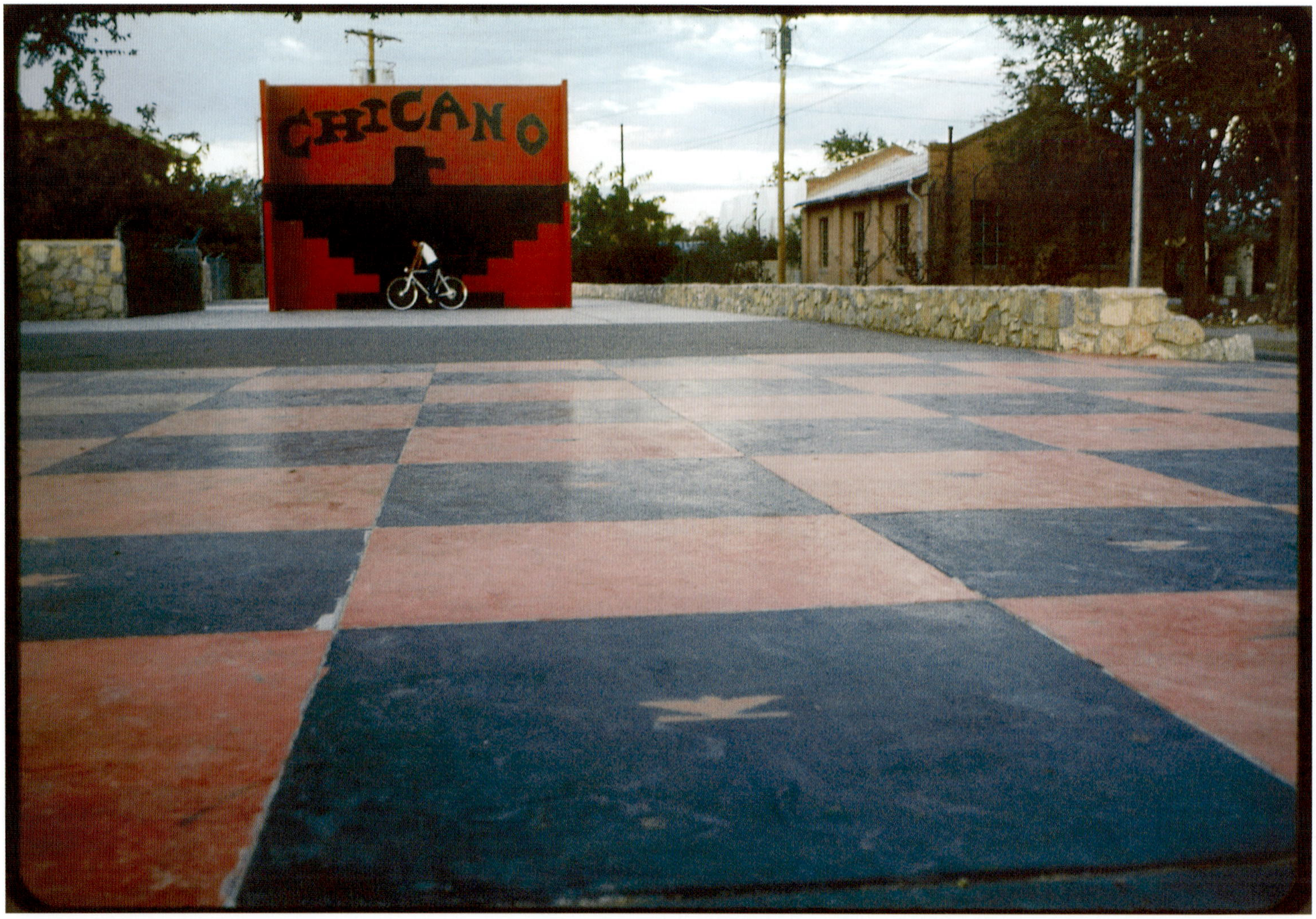

"Thunderbird display in the Spanish speaking Second Ward."

Danny Lyon, El Paso, Texas, July 1972

(412-DA-2830)

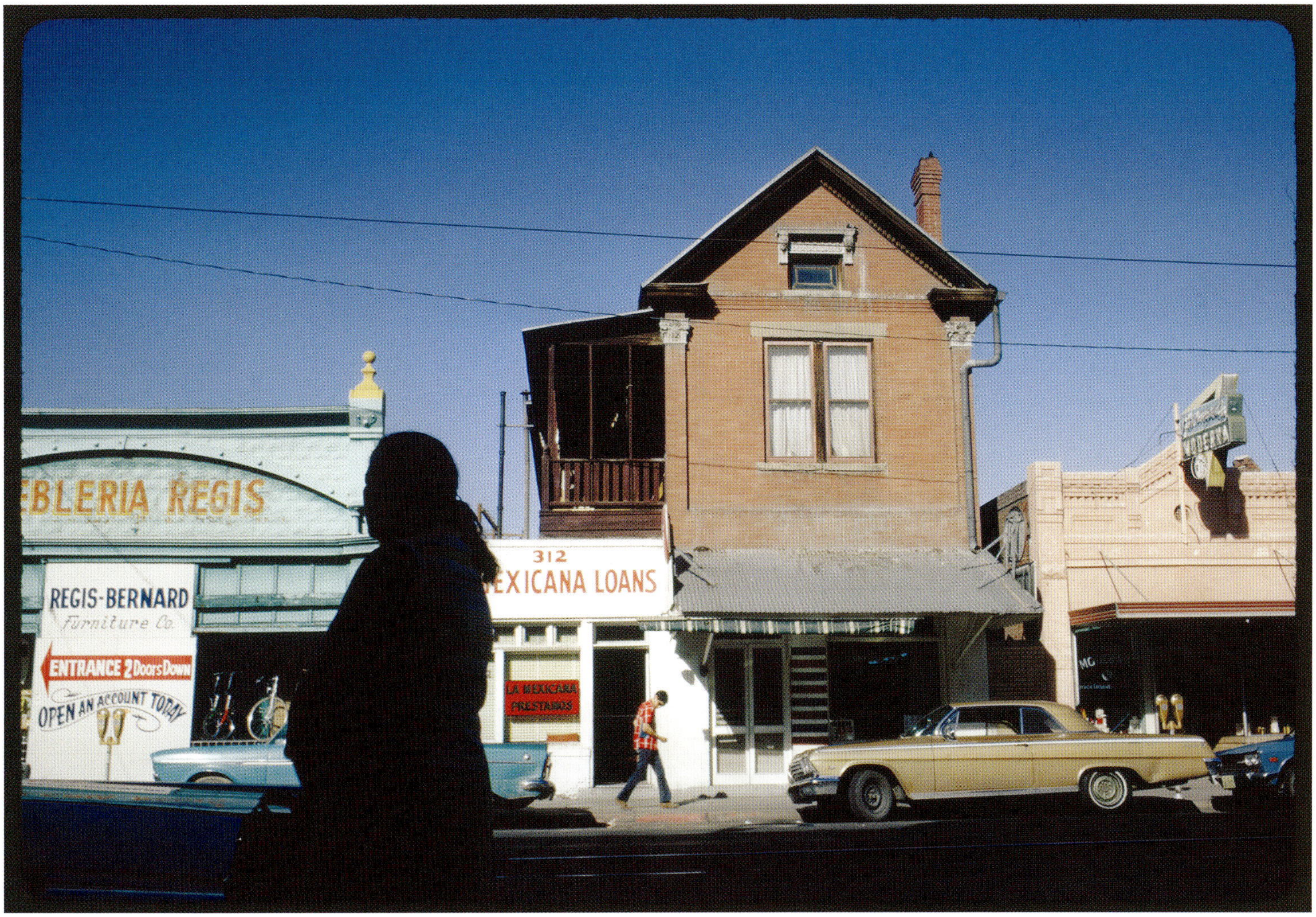

"Stanton Street in the second ward, the Spanish-speaking section."

Danny Lyon, El Paso, Texas, June 1972

(412-DA-2840)

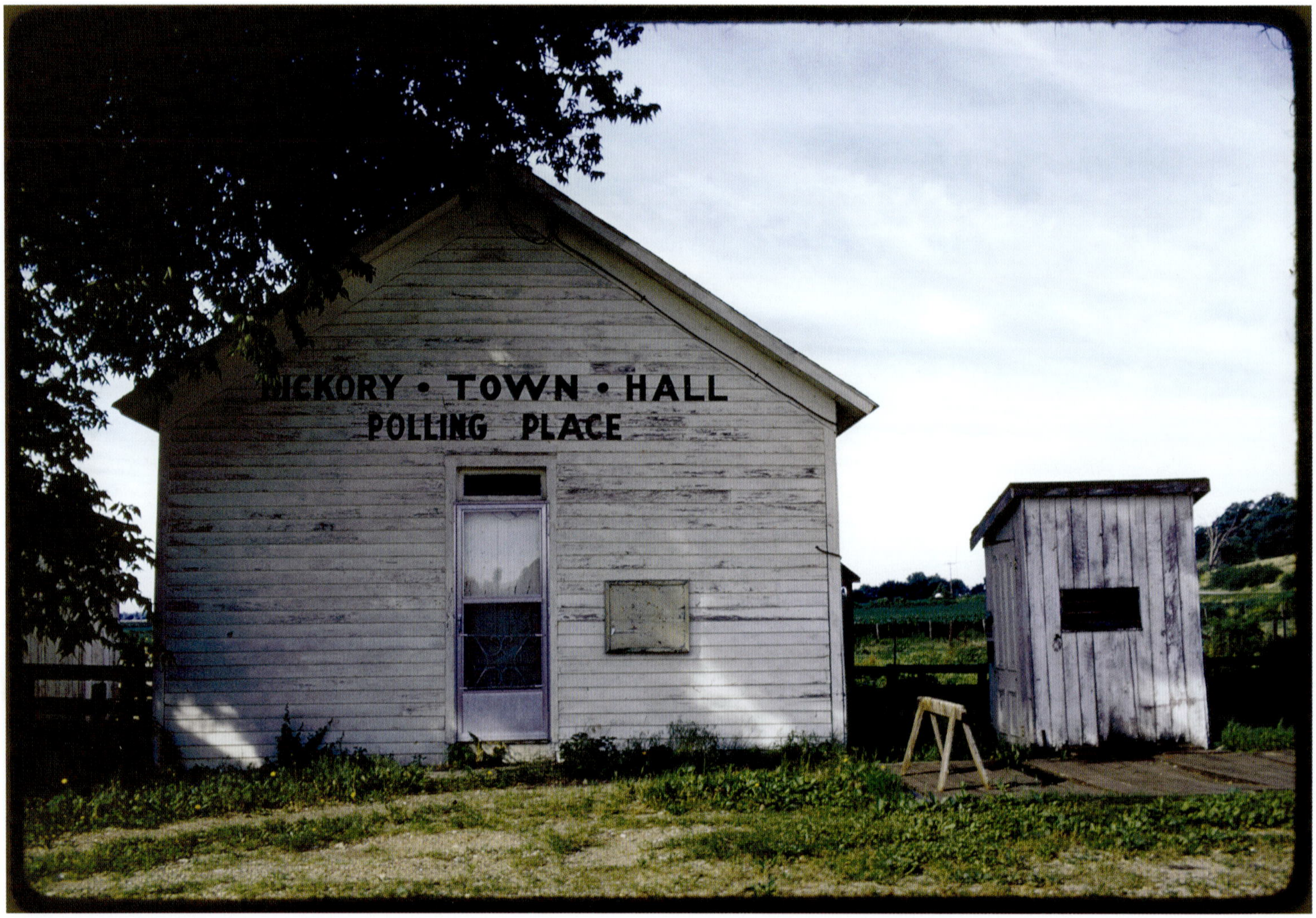

"Hickory Town Hall and polling place."

Arthur Greenberg, Hickory, Illinois, June 1973

(412-DA-10024)

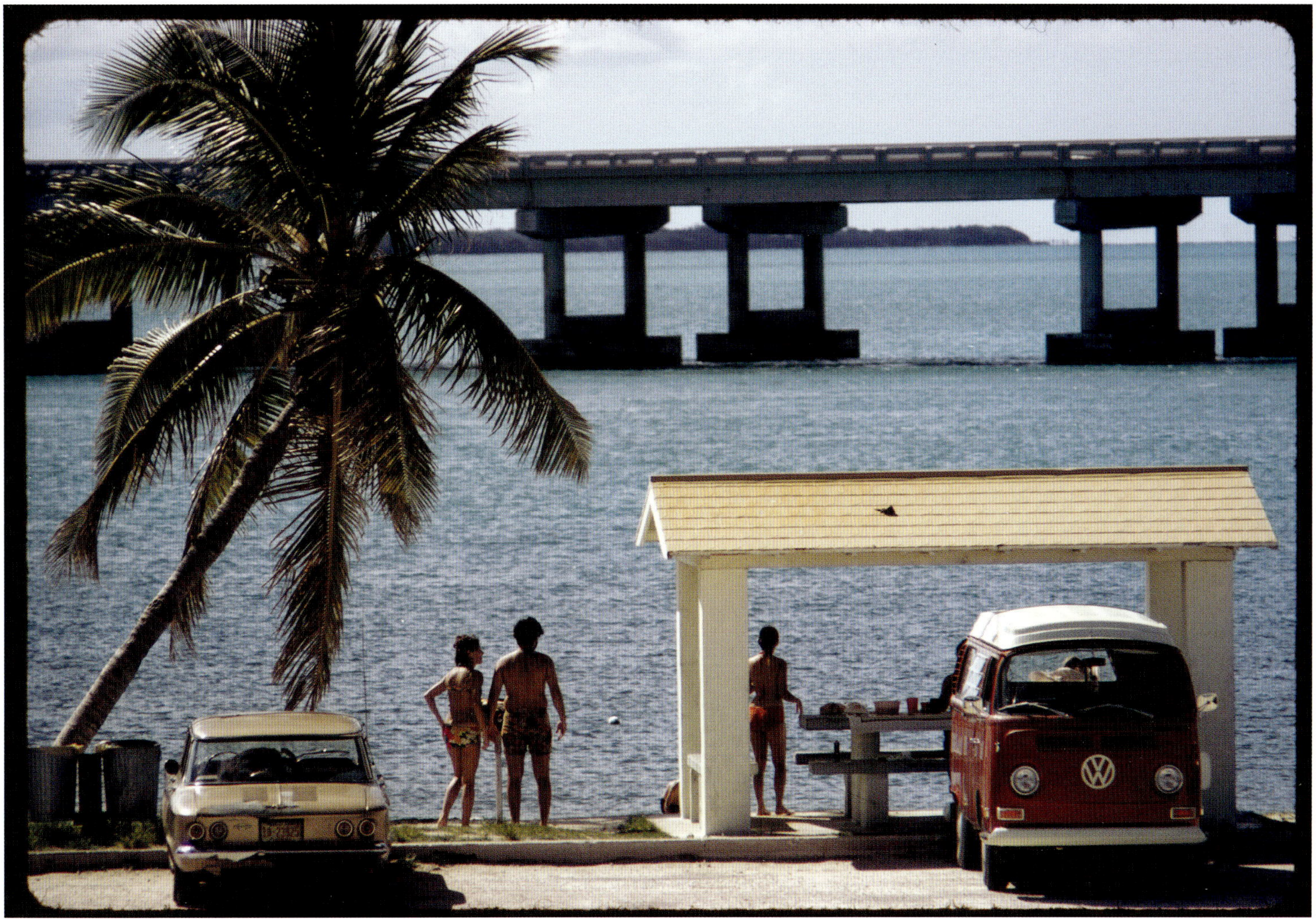

"At Bahia Honda State Park, on Bahia Honda Key, Florida."

Flip Schulke, Bahia Honda Key, Florida, June 1973

(412-DA-6317)

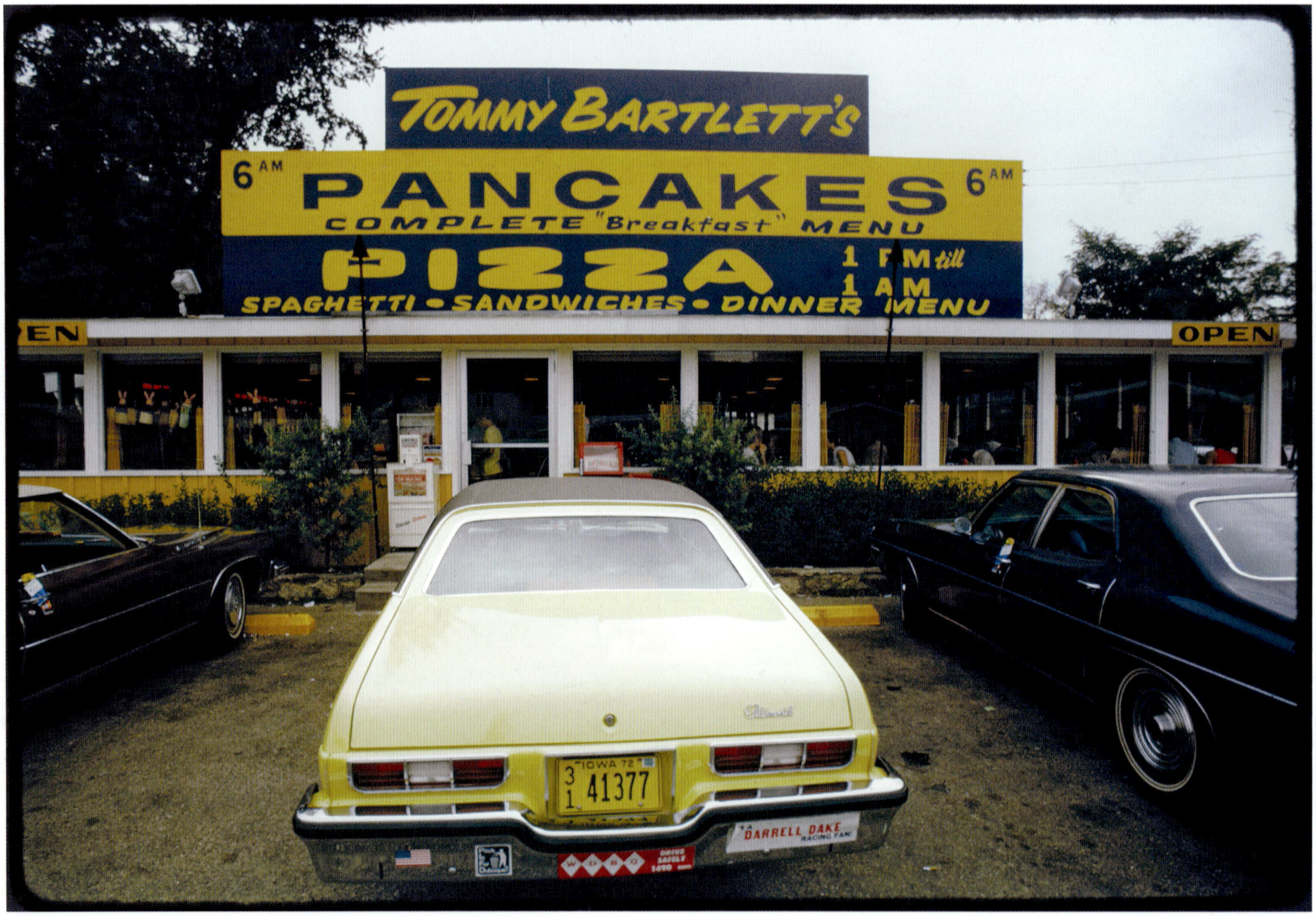

"Near the town of Wisconsin Dells the Wisconsin River channels through deep, soft sandstone cliffs, cutting rock into fantastic shapes. These natural splendors have given rise to a booming tourist industry. People come in droves, often in campers and trailers. Boat trips, shops, bars, and diversions of every kind vie for patronage in an amusement complex extending 2 or 3 miles beyond the town."

Jonas Dovydenas, Wisconsin Dells, Wisconsin, September 1973

(412-DA-11736)

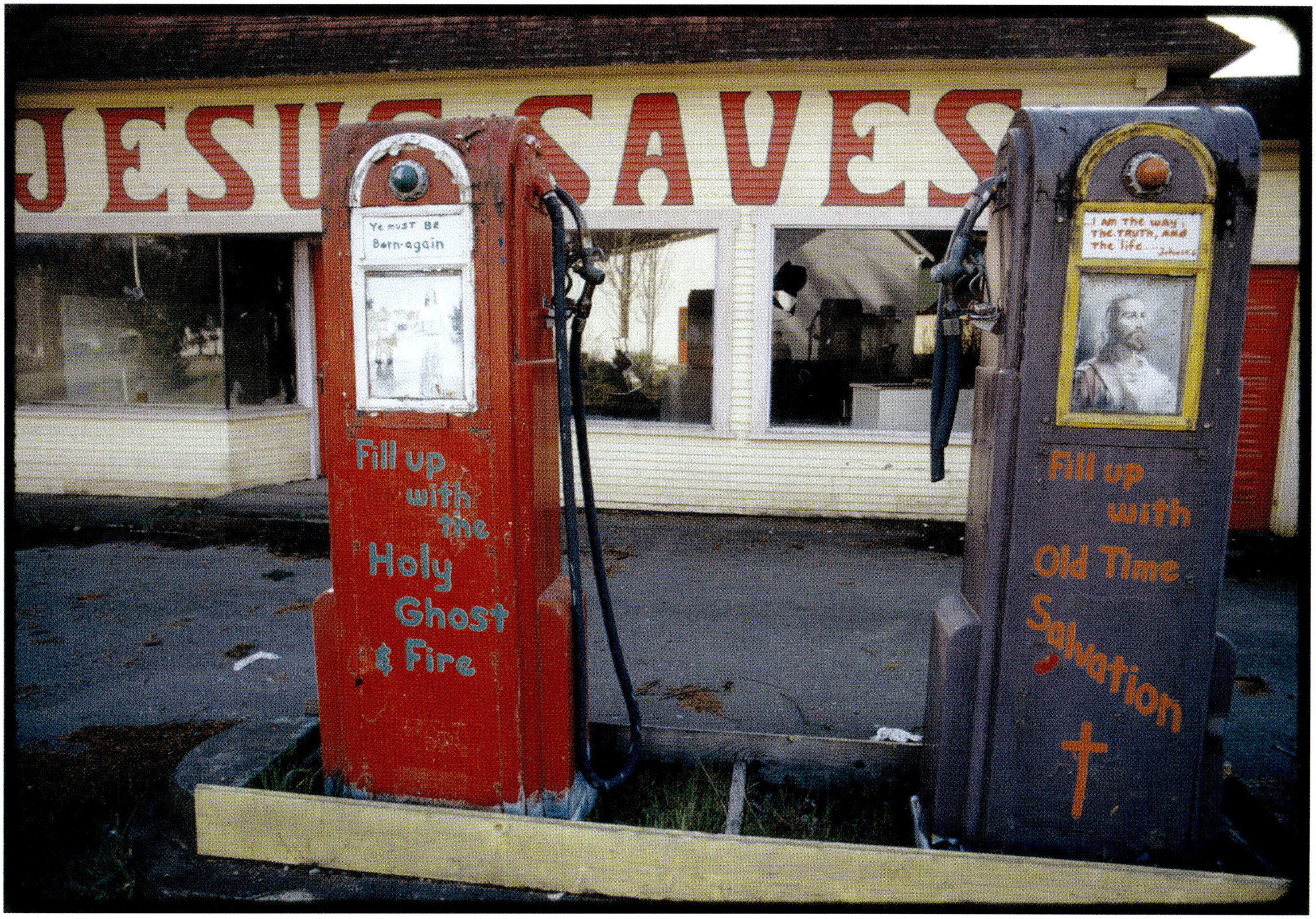

"Gasoline stations abandoned during the fuel crisis in winter of 1973–74 were sometimes used for other purposes. This station at Potlatch, Washington, west of Olympia was turned into a religious meeting hall. Signs painted on the gas pumps proclaim 'Fill up with the Holy Ghost . . . and Salvation.'"

David Falconer, Potlatch, Washington, April 1974

(412-DA-13061)

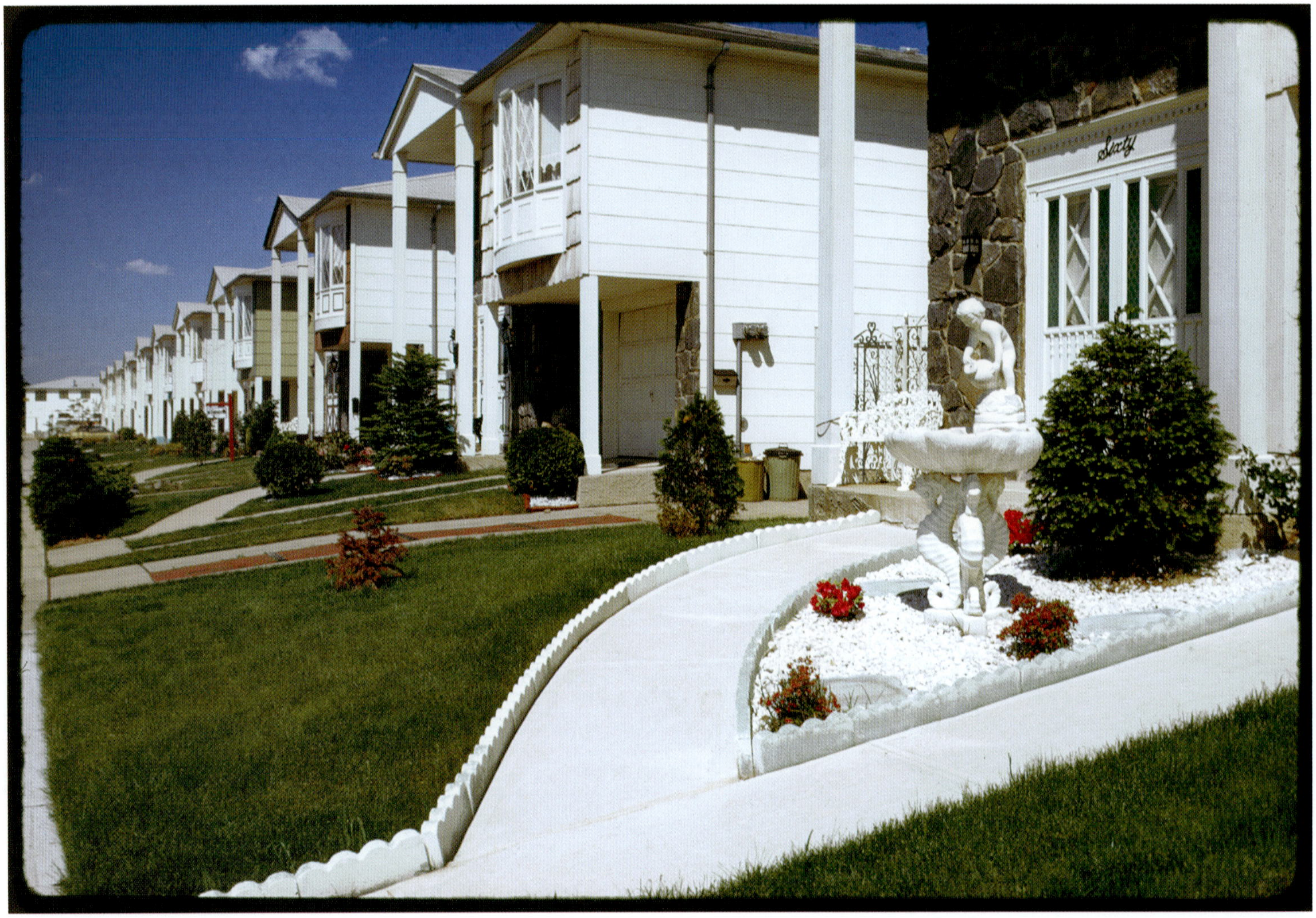

"Great Kills Park, Staten Island."

Arthur Tress, Staten Island, New York, May 1973

(412-DA-11375)

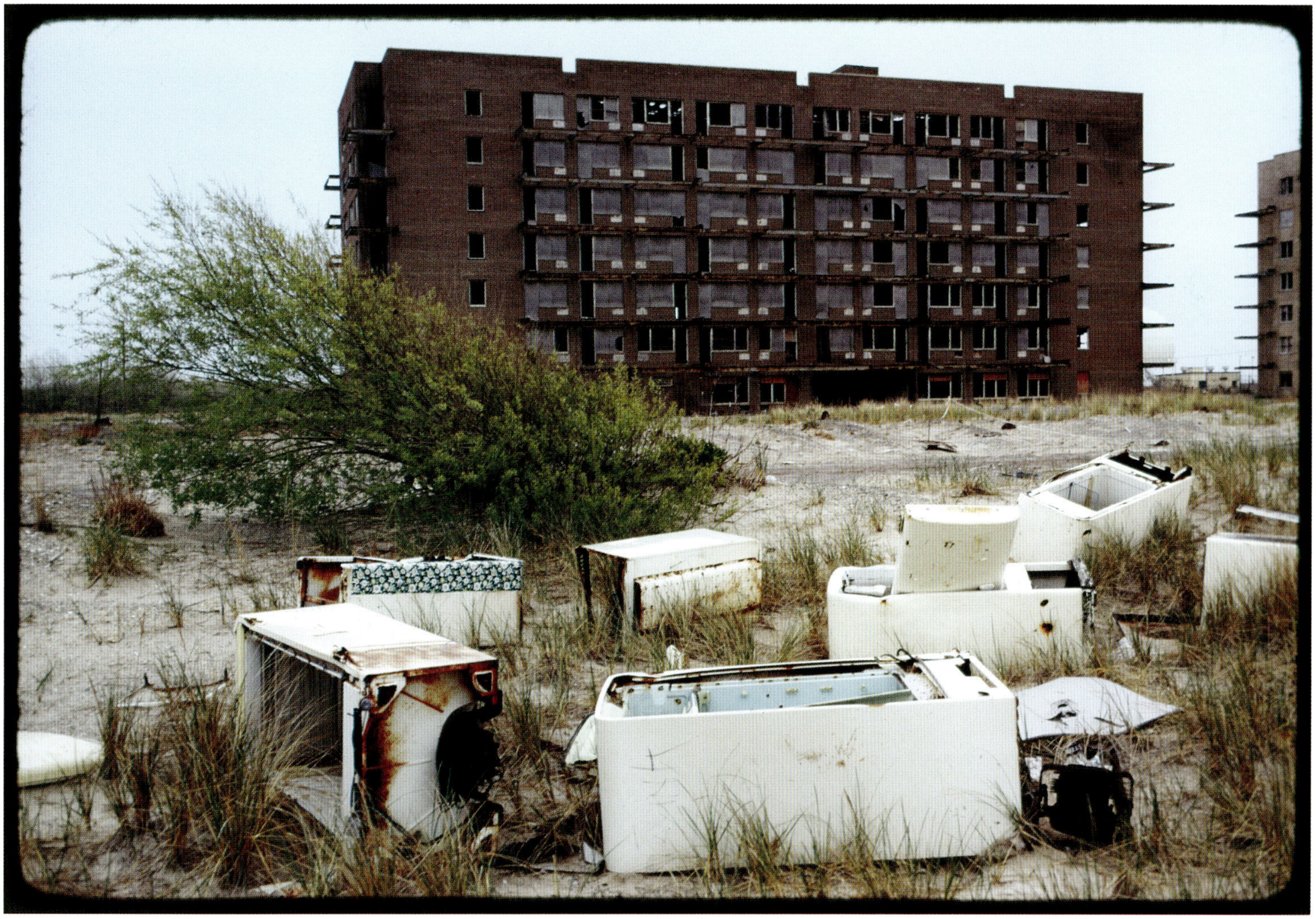

"Breezy Point, Long Island."

Arthur Tress, Long Island, New York, May 1973

(412-DA-11410)

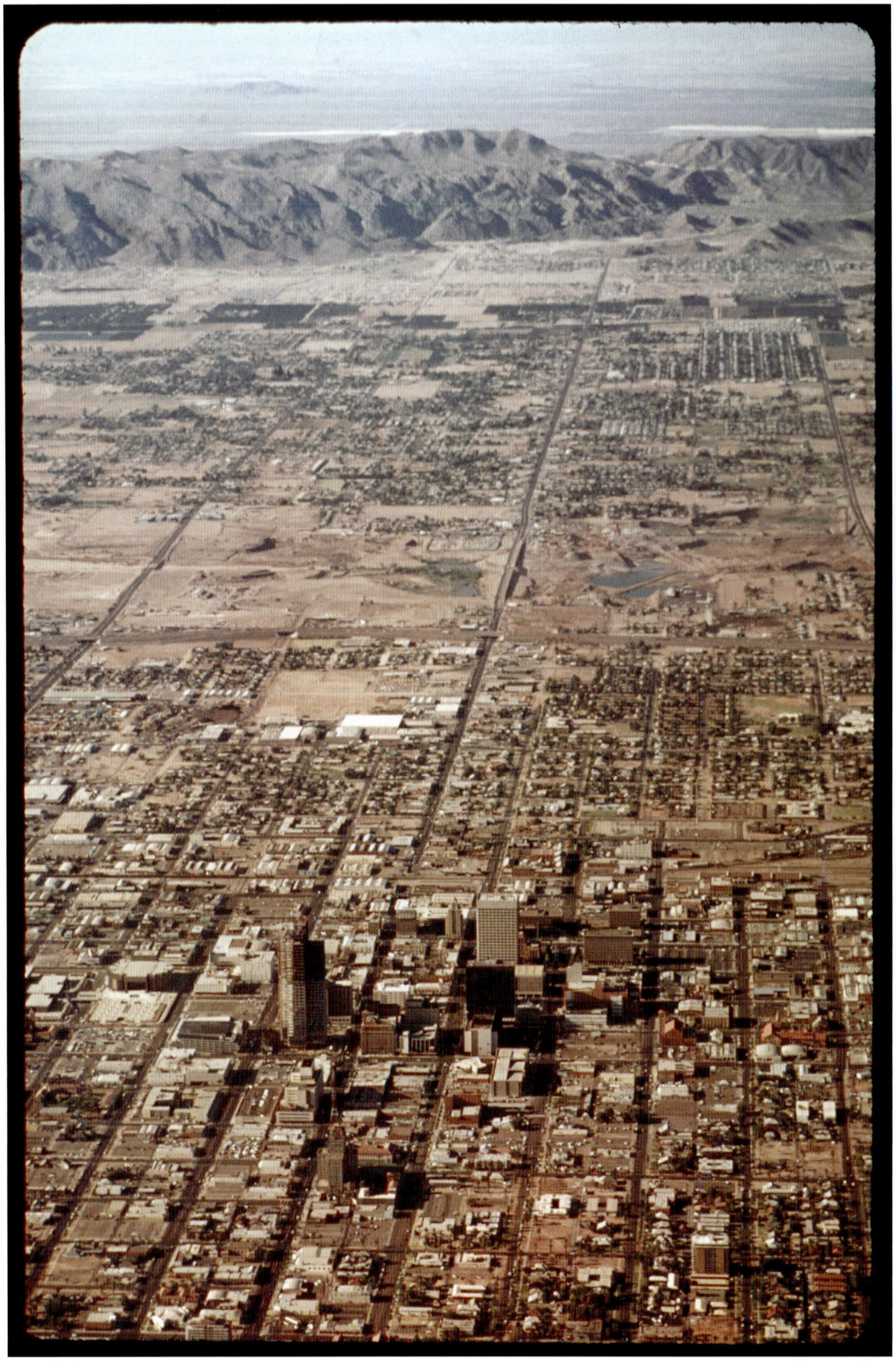

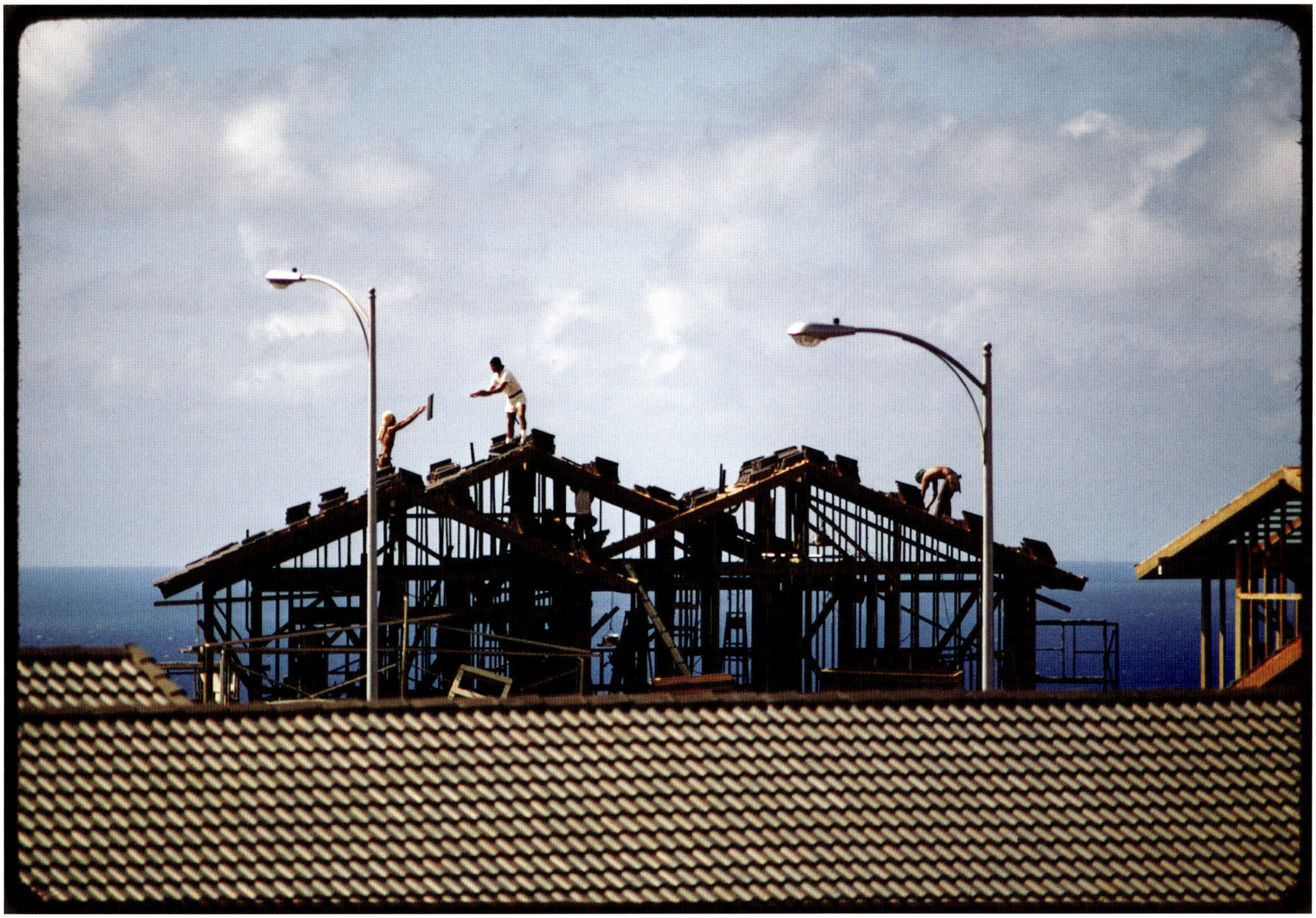

“Hawaii-Kai, built on a swamp and a hillside, is a fast-growing area east of downtown Honolulu, scarcity of land and skyrocketing costs are favoring construction of town houses and high-rise apartments over the once dominant single-family unit.”

Charles O’Rear, Honolulu, Hawaii, October 1973

(412-DA-11472)

Opposite: “The view of Phoenix’s urban sprawl from 4000 ft. South Mountain in the background.”

Cornelius M. Keyes, Phoenix, Arizona, June 1972

(412-DA-1580)

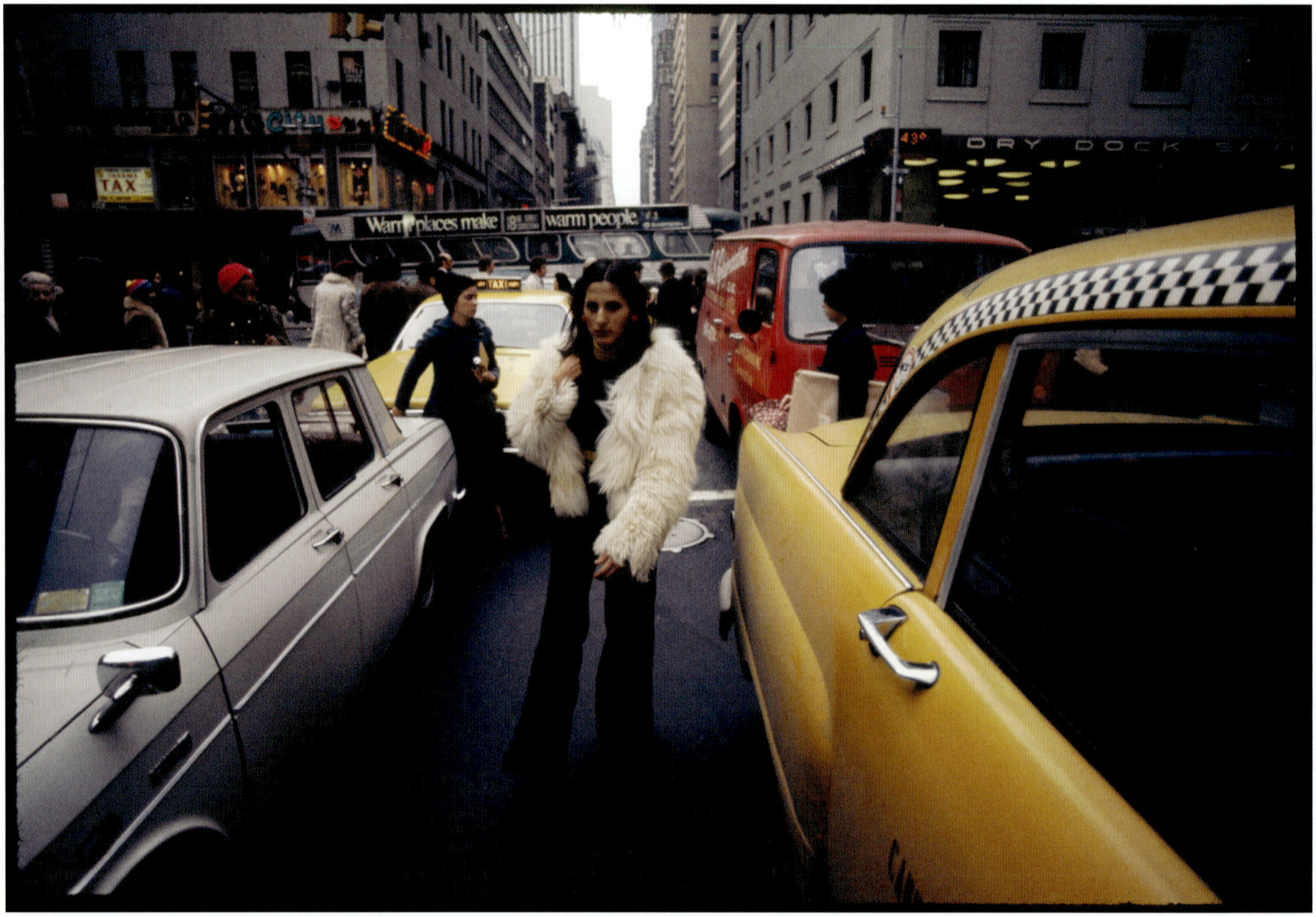

"Midtown traffic congestion and jaywalking pedestrians."

Dan McCoy, New York, New York, April 1973

(412-DA-7378)

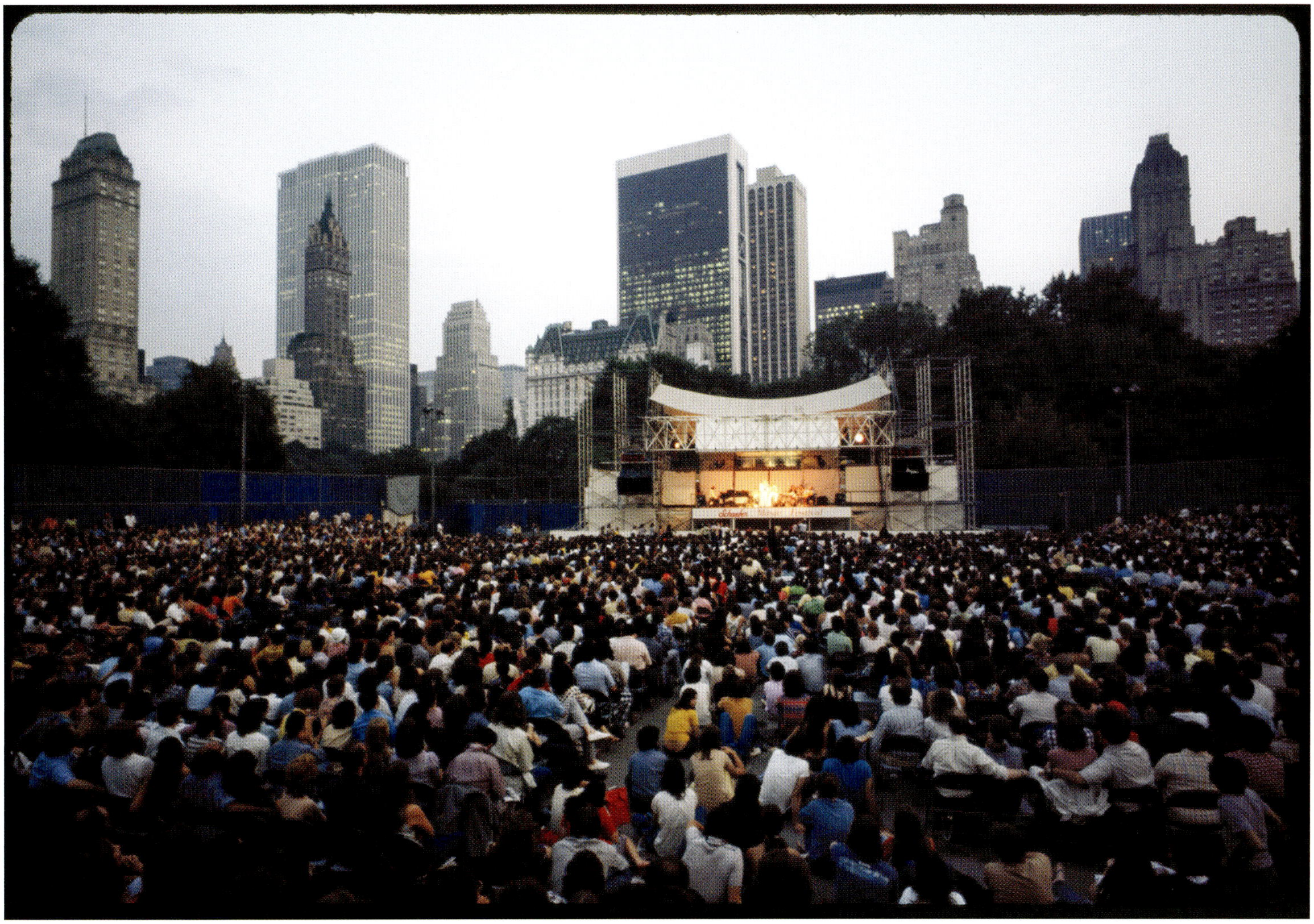

"Crowd gathered at the Schaefer Bandstand in Central Park to hear singer Judy Collins also get a dramatic view of the Towers of Midtown Manhattan."

Suzanne Szasz, New York, New York, June 1973

(412-DA-9188)

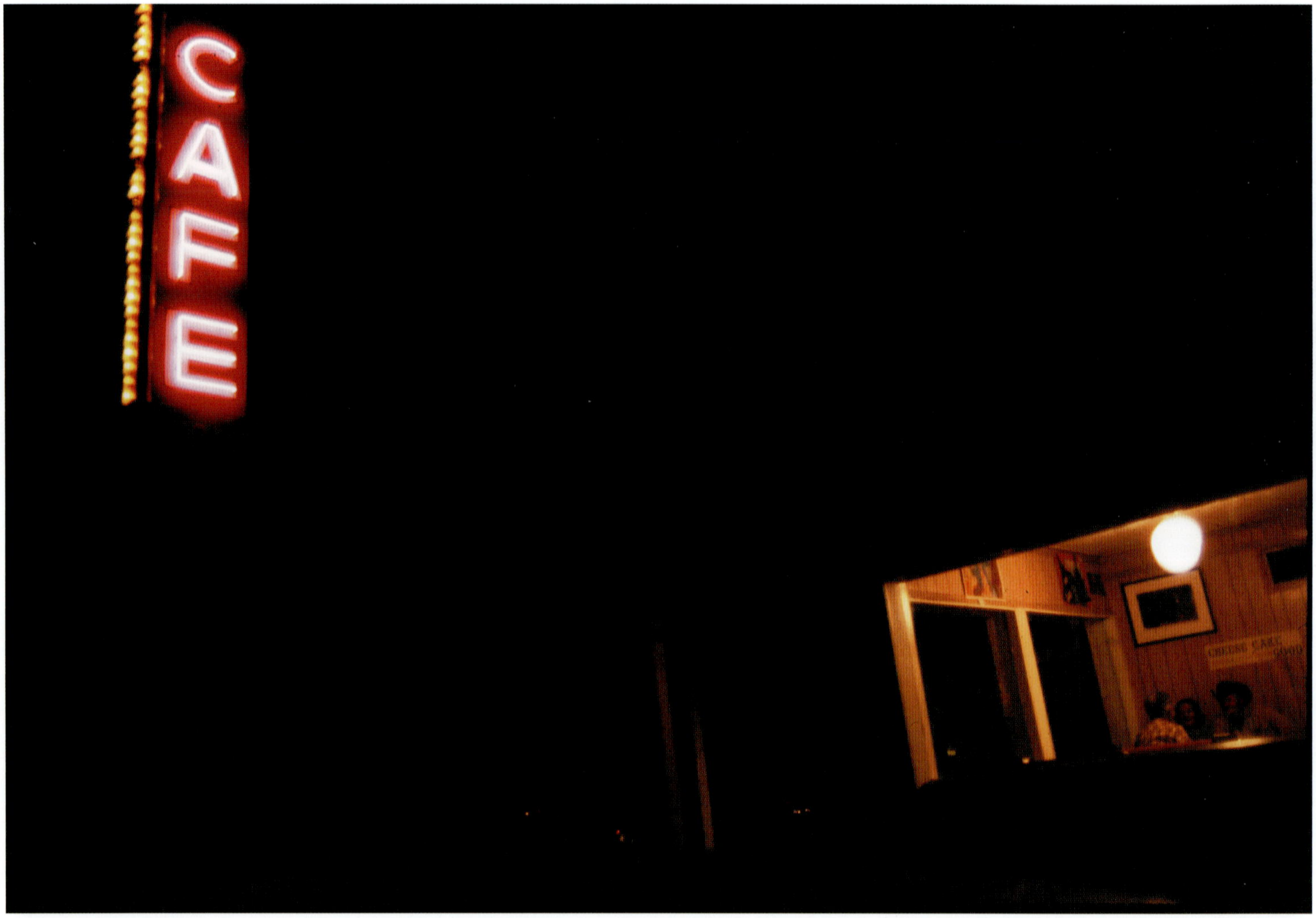

"Ranchers having coffee."

Tomas Sennett, Laytonville, California, May 1972

(412-DA-401)

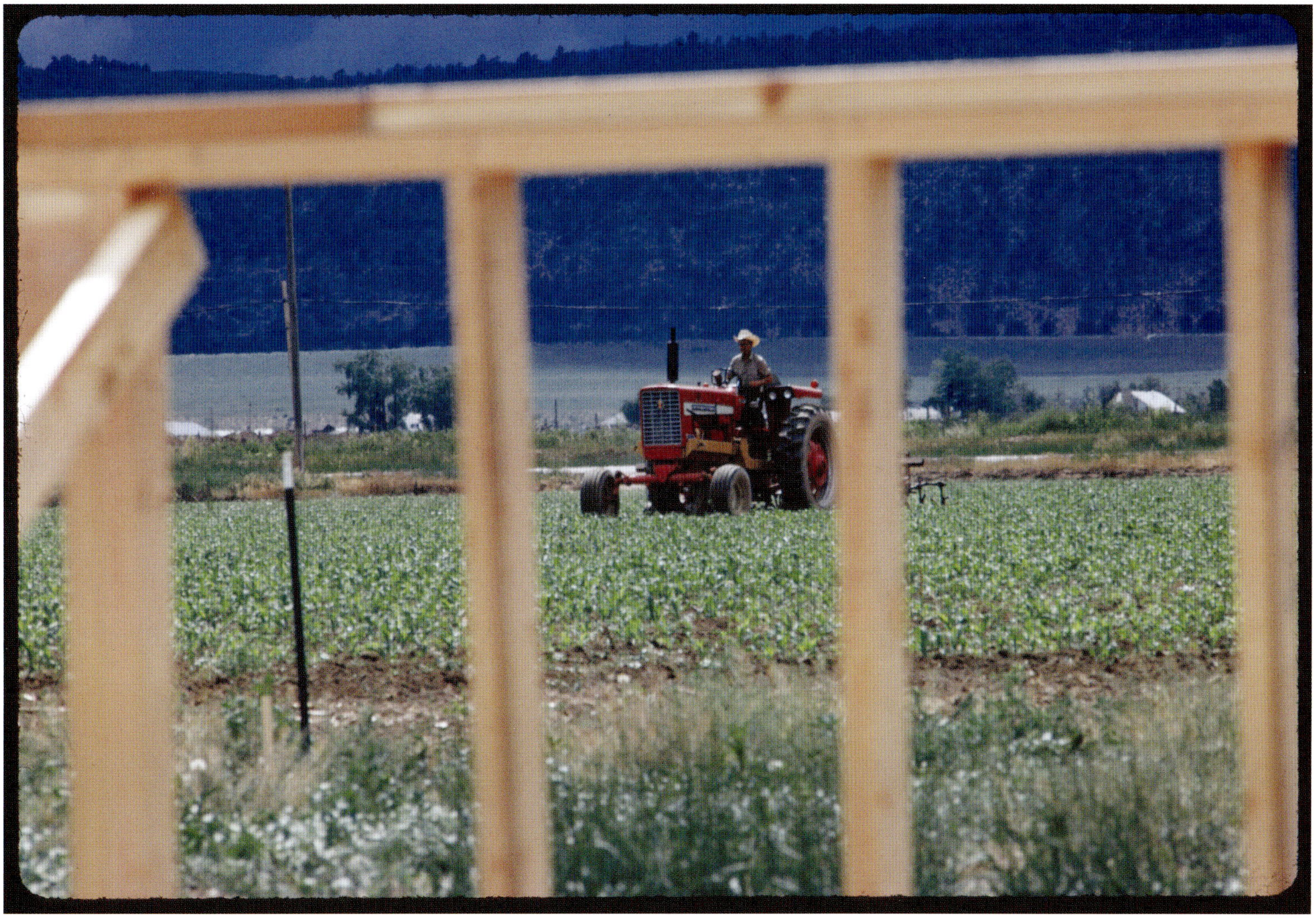

"Housing development encroaches upon farm land."

Bill Gillette, Fort Collins, Colorado, June 1972

(412-DA-1174)

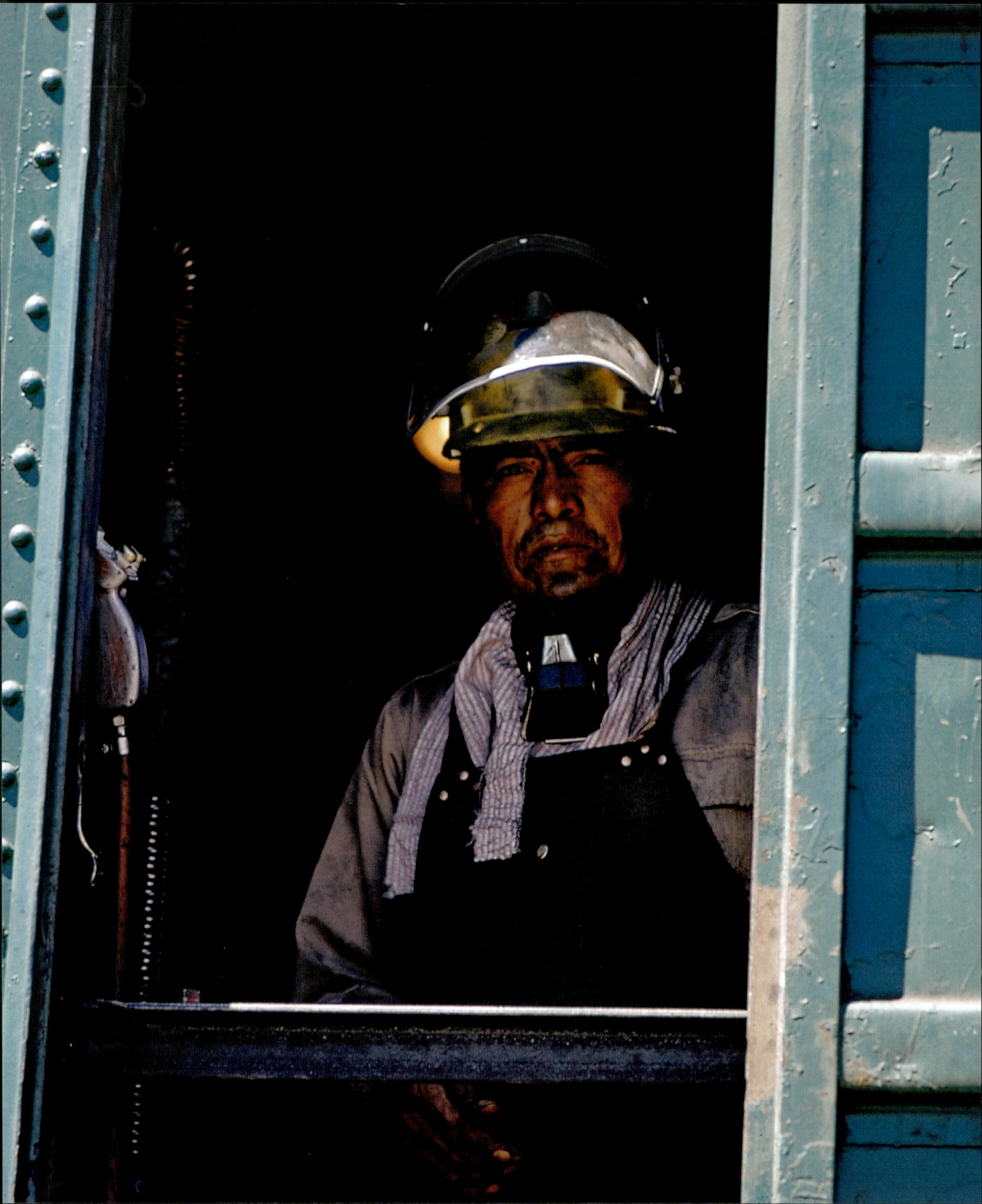

On Assignment: Lyntha Scott Eiler's Arizona

In spring 1972, Lyntha Scott Eiler and her husband, Terry Eiler, undertook a DOCUMERICA assignment photographing northern Arizona. Lyntha's photographs touched on a variety of subjects—encroaching development, tourism, Native American children, strip mining, and the building of the Navajo Generating Station. In addition to "the awe of the Grand Canyon, the high desert plateaus and the cultures of Native Americans and pioneers," she witnessed "the enormous draglines on the sacred Black Mesa, the Four Corners power plant's plume . . . and the depleting of the pristine Navajo aquifer." Her photographs aimed "to document my own era," as well as "to follow in the steps of [FSA photographers] Marion Post Wolcott, Dorothea Lange, and Ester Bubley."

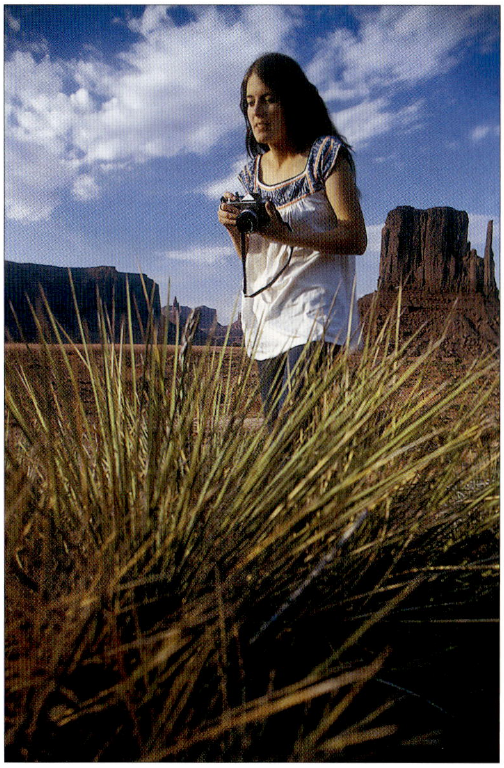

Lyntha Scott Eiler, Monument Valley, Arizona, 1973
Photograph by Terry Eiler
Courtesy of Terry Eiler

Today, Lyntha Scott Eiler works as a documentary photographer as well as a photo editor and researcher. She has taught at Ohio University, Marietta College, and the Salt Institute for Documentary Studies. Her work has appeared in *Ohio Magazine*, *Sunset*, and *National Geographic*. In addition to the National Archives, her photographs are held at the Library of Congress, the Museum of Northern Arizona, and the Calvert Marine Museum in Maryland. She has published books on the Blue Ridge Mountains, high school basketball, and the Havasupai Indians.

Opposite: "A Navajo construction worker at the Navajo Generating Plant. When completed, this will be the largest such plant in Arizona."

Lyntha Scott Eiler, Page, Arizona, May 1972 (detail, see page 117)

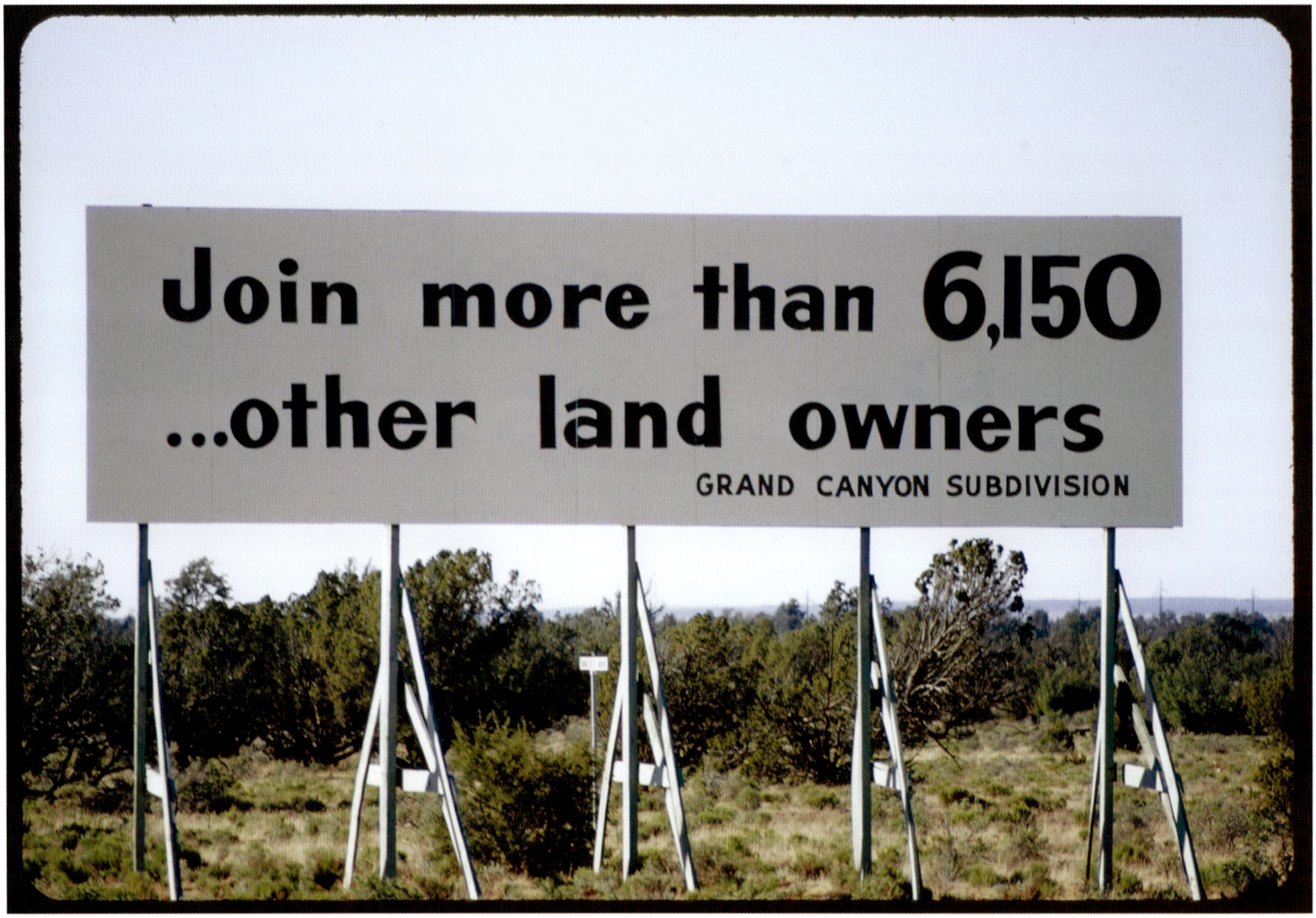

"Grand Canyon Estates."

Lyntha Scott Eiler, Arizona, June 1972

(412-DA-1643)

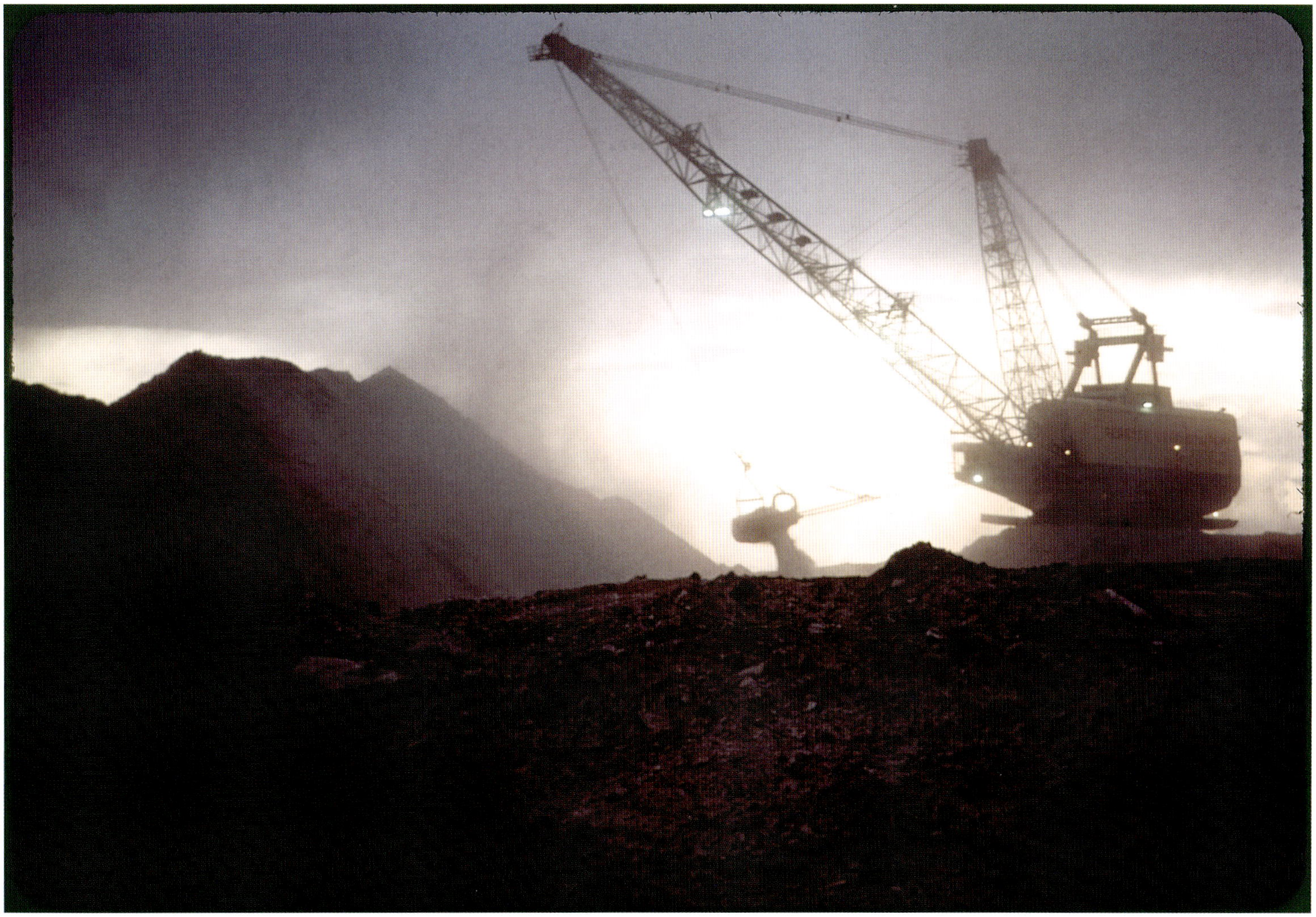

"Strip Mining on Indian burial grounds by Peabody Coal Co."

Lyntha Scott Eiler, Black Mesa, Arizona, May 1972

(412-DA-1616)

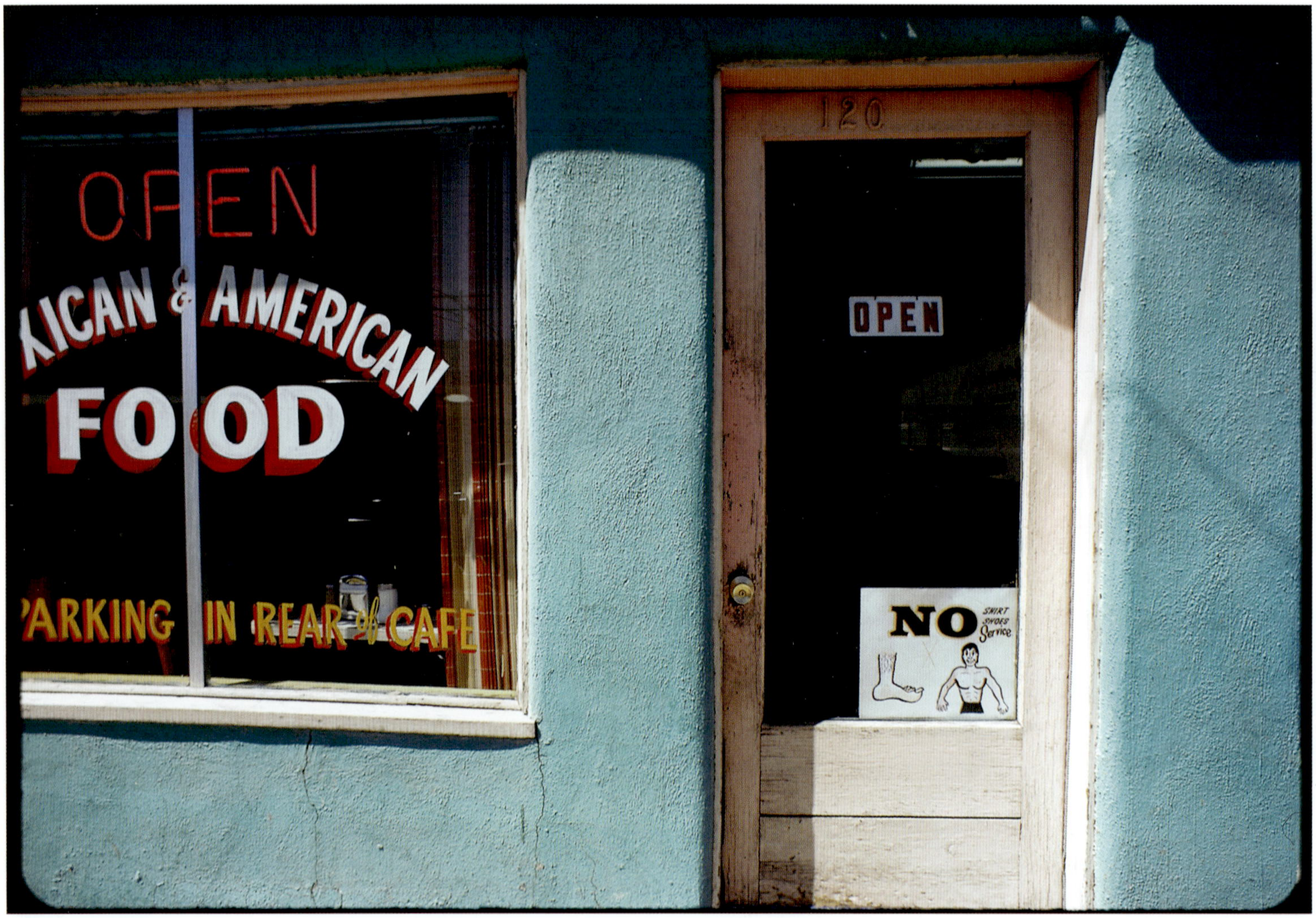

"Arizona—Holbrook"

Lyntha Scott Eiler, Holbrook, Arizona, June 1972

(412-DA-1651)

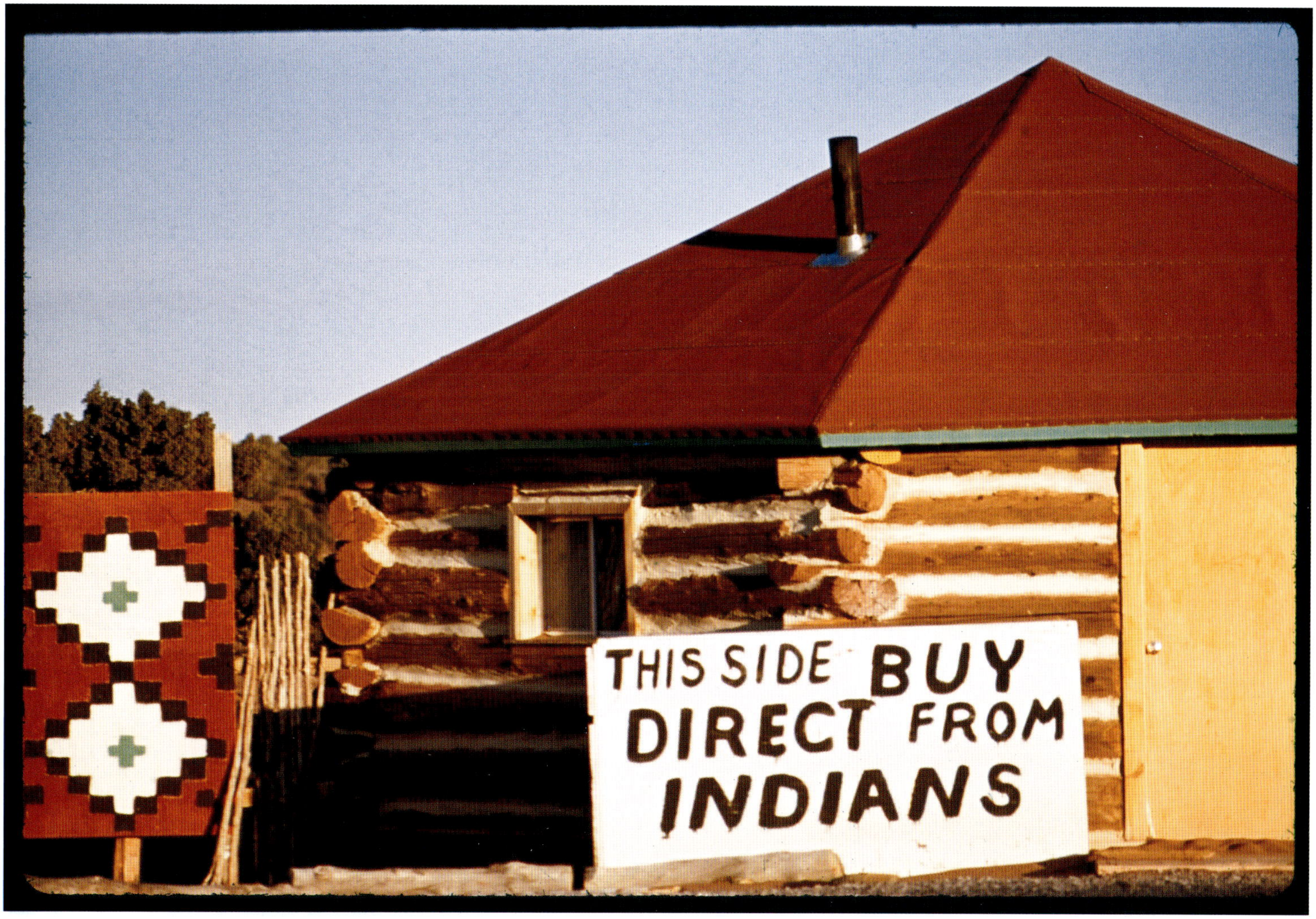

"Indian-owned shop on one side of Route 66 challenges white-owned shop on the other."

Lyntha Scott Eiler, Arizona, June 1972

(412-DA-1655)

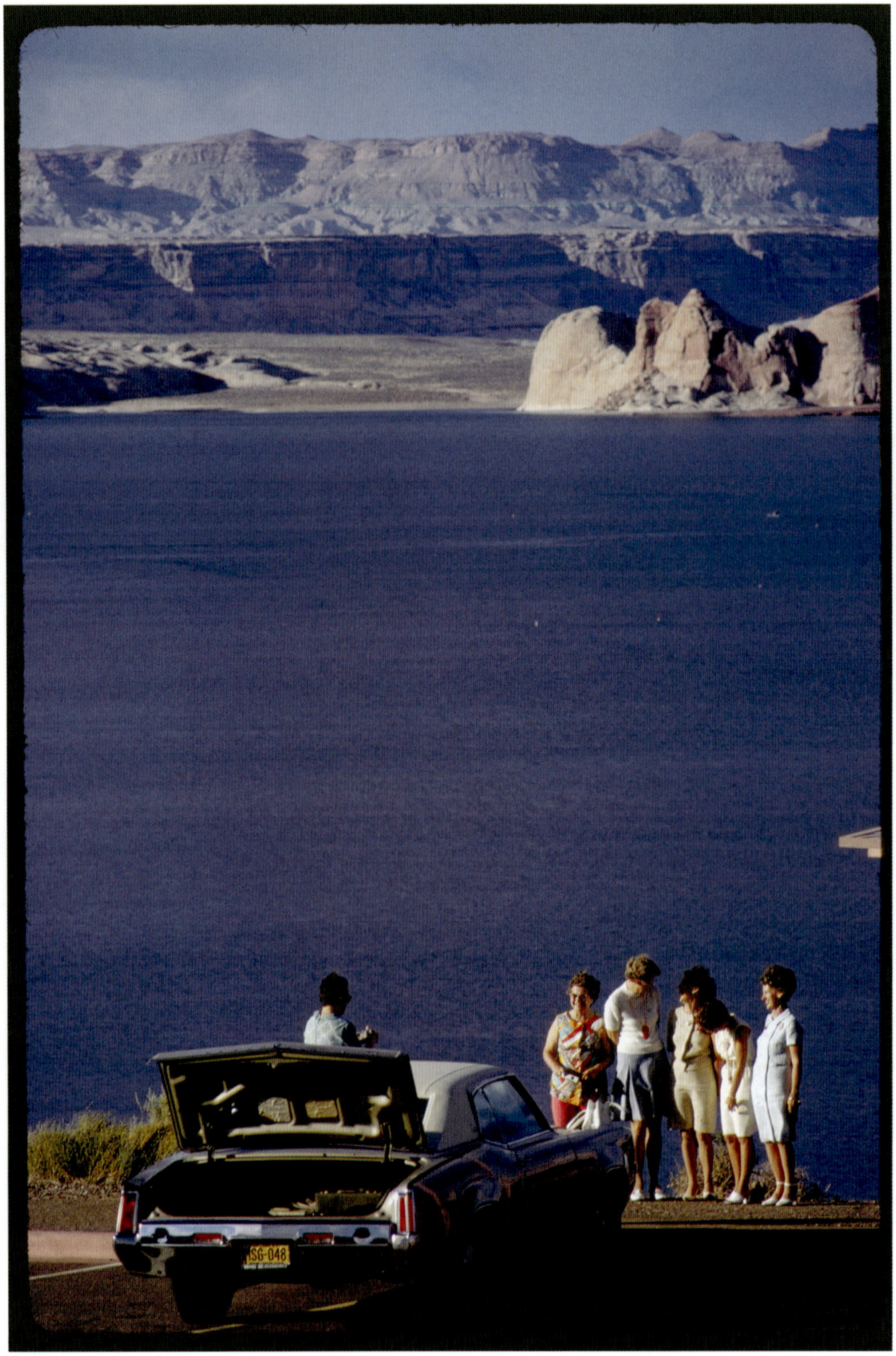

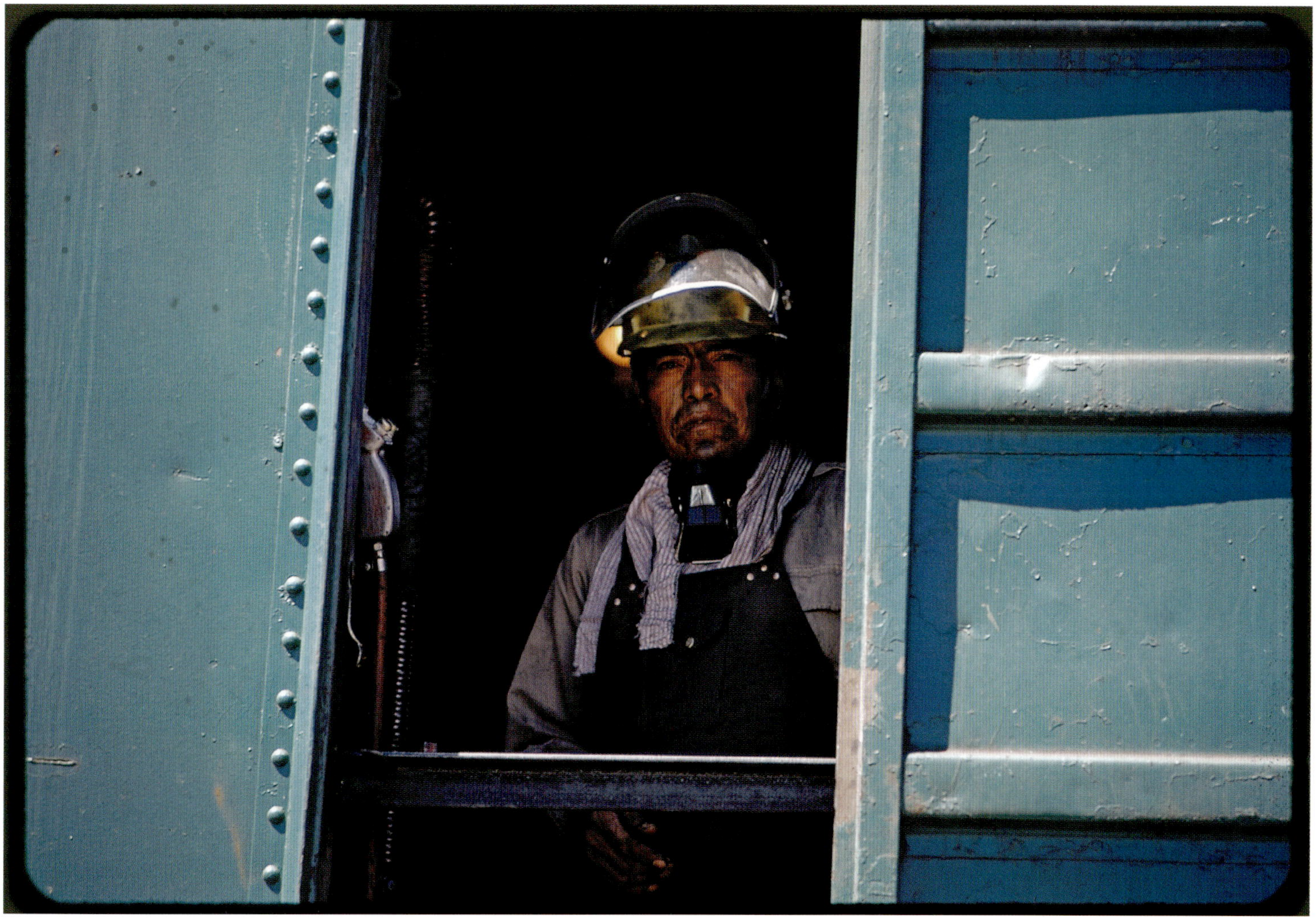

"A Navajo construction worker at the Navajo Generating Plant. When completed, this will be the largest such plant in Arizona."

Lyntha Scott Eiler, Page, Arizona, May 1972

(412-DA-1670)

Opposite: "Arizona—near Page"

Lyntha Scott Eiler, Page, Arizona, May 1972

(412-DA-1896)

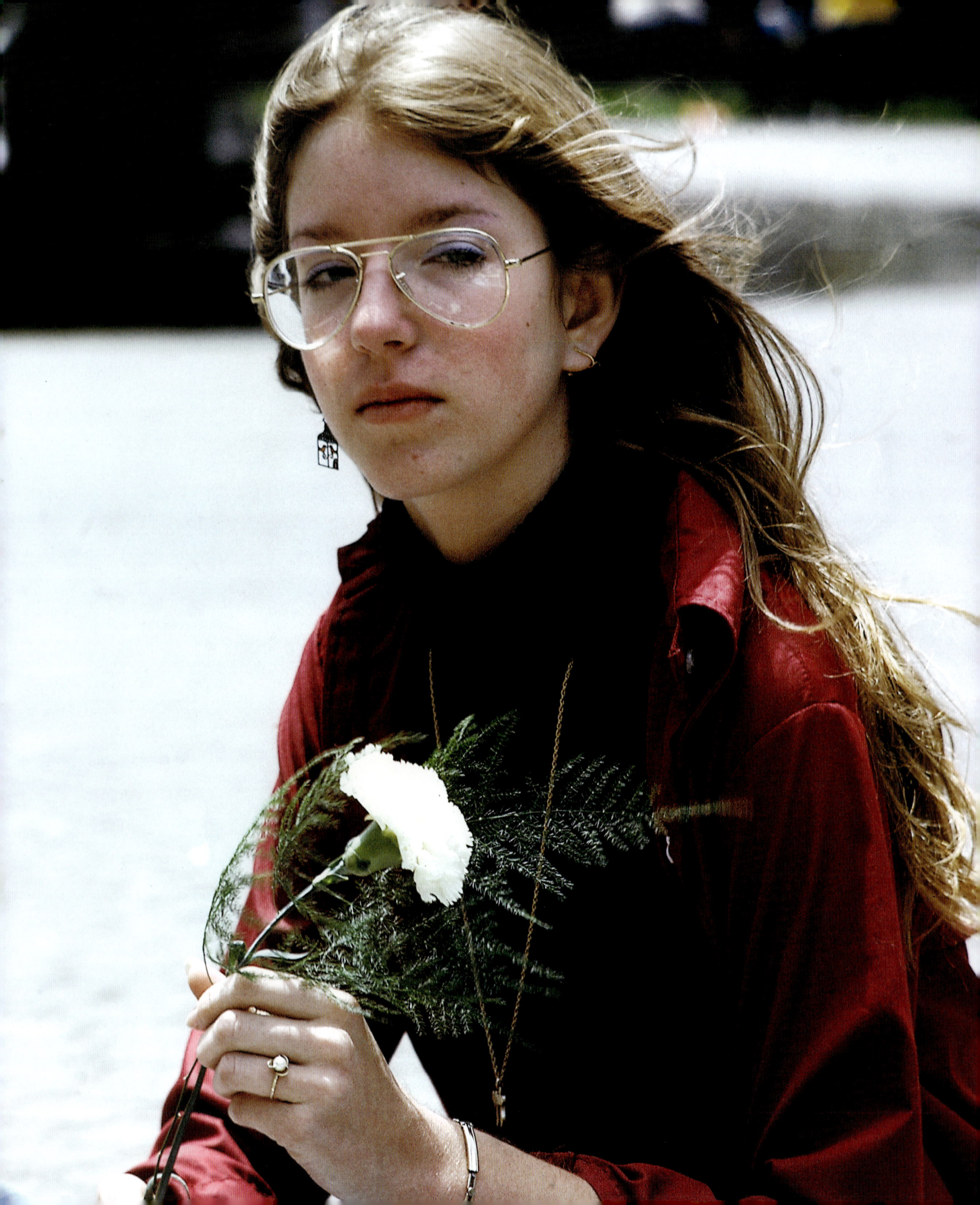

On Assignment: Tom Hubbard's Fountain Square

Tom Hubbard's DOCUMERICA assignment was in a place familiar to him—Fountain Square in downtown Cincinnati, Ohio. Hubbard had covered many events there during his years as a photojournalist with the *Cincinnati Enquirer*, but in 1973, he staked out the square for several months. He recorded festivals, political demonstrations, and concerts, as well as the comings and goings of city life. Hubbard photographed chess matches, kite flying, office workers at lunch, and street scenes. His shots especially captured the changes in dress and personal appearance emerging in 1970s America.

Hubbard's Fountain Square assignment profoundly affected how he thought about photography. He moved away from photojournalism and the "forced timing" of "staged" news stories and toward documentary photography that emphasized the ordinary and the "non-event." He said that photographing Fountain Square gave him "a profound respect for photography's ability to record the full range of the human scene."

In addition to his photographs for the *Enquirer*, Tom Hubbard's work has appeared in *Time*, *People*, *US Magazine*, the *New York Times*, *Business Week*, *Fortune*, and *Der Spiegel*. He taught photojournalism at The Ohio State University from 1993 to 1998. Since 2000 he has been creating abstract digital art.

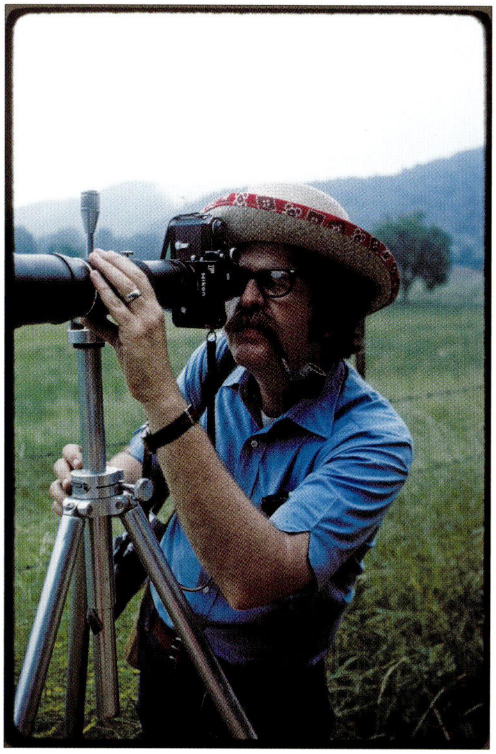

Tom Hubbard, 1975
Photograph by Ingrid Hubbard
Courtesy of Tom Hubbard

National Archives, Records of the Environmental Protection Agency

(412-DM-0148-7-14)

Opposite: "Fountain Square in downtown Cincinnati is a public square that works for the city and its people in a myriad of ways: a flower vender has made a sale,"

Tom Hubbard, Cincinnati, Ohio, May 1973
(detail, see page 122)

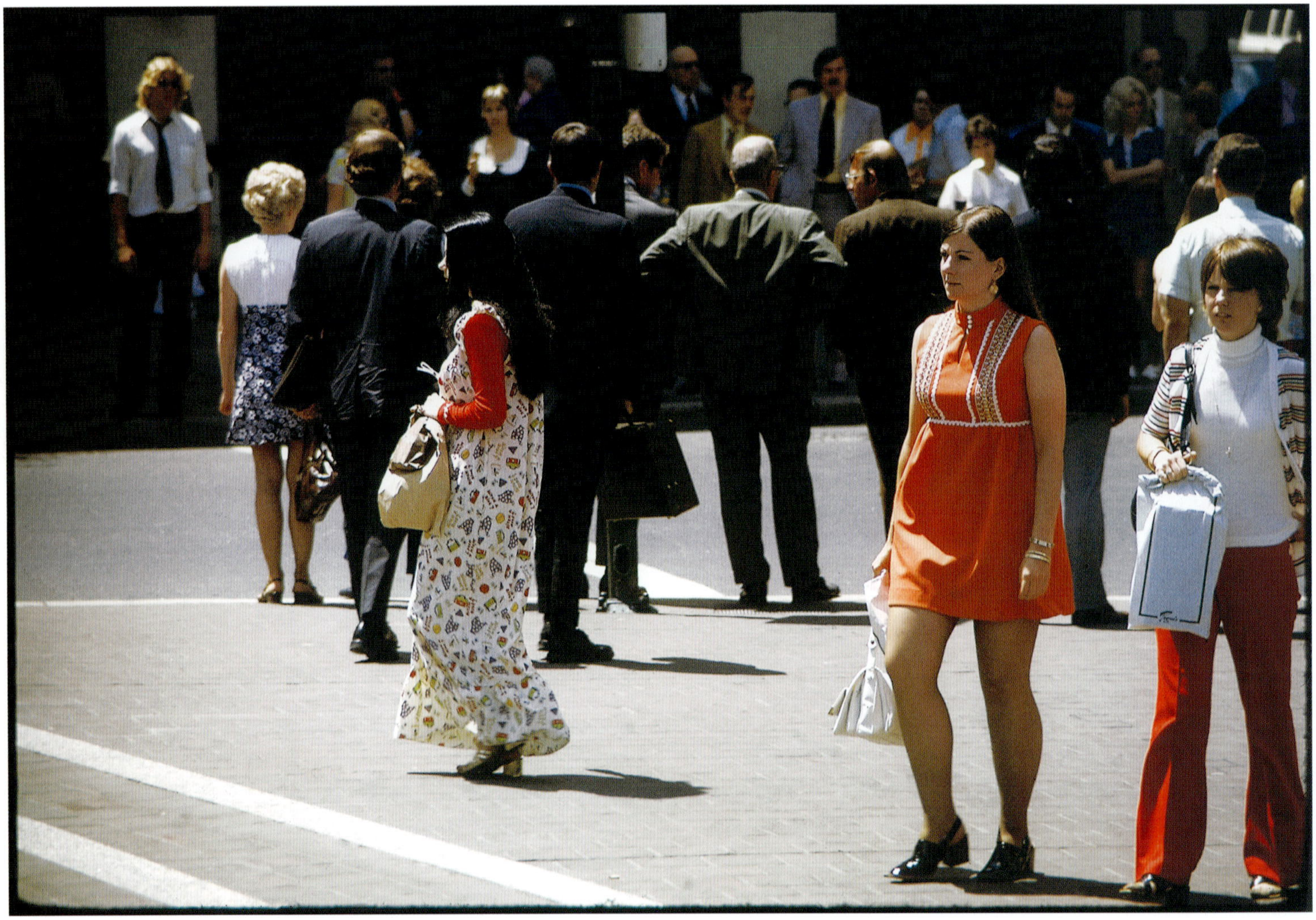

"Fountain Square in downtown Cincinnati is a public square that works for the city and its people in a myriad of ways: At corner of 5th and Vine Streets."

Tom Hubbard, Cincinnati, Ohio, June 1973

(412-DA-10725)

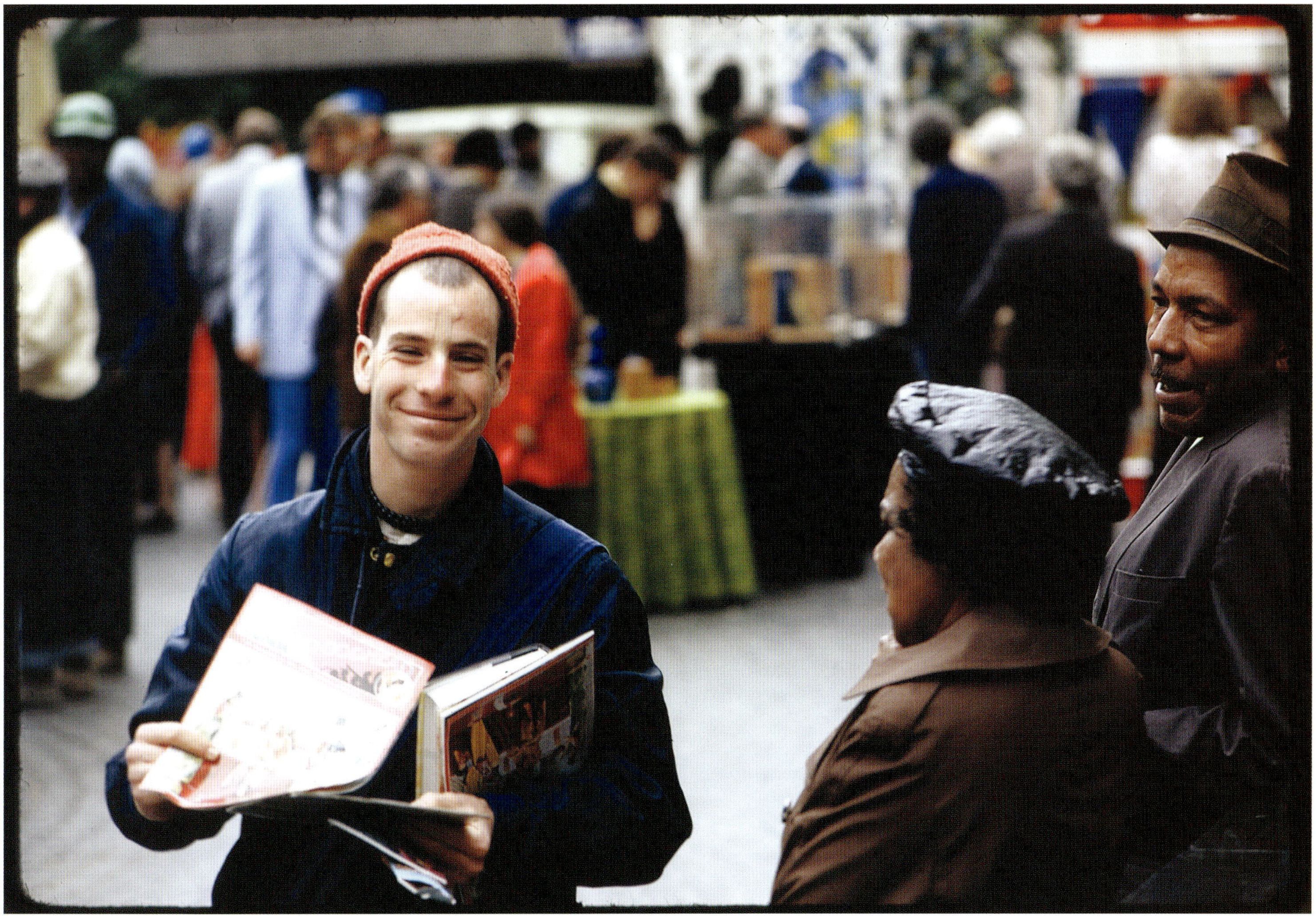

"Fountain Square in downtown Cincinnati is a public square that works for the city and its people in a myriad of ways: Distributing Hare Krishna literature at the Israeli birthday celebration."

Tom Hubbard, Cincinnati, Ohio, May 1973

(412-DA-10682)

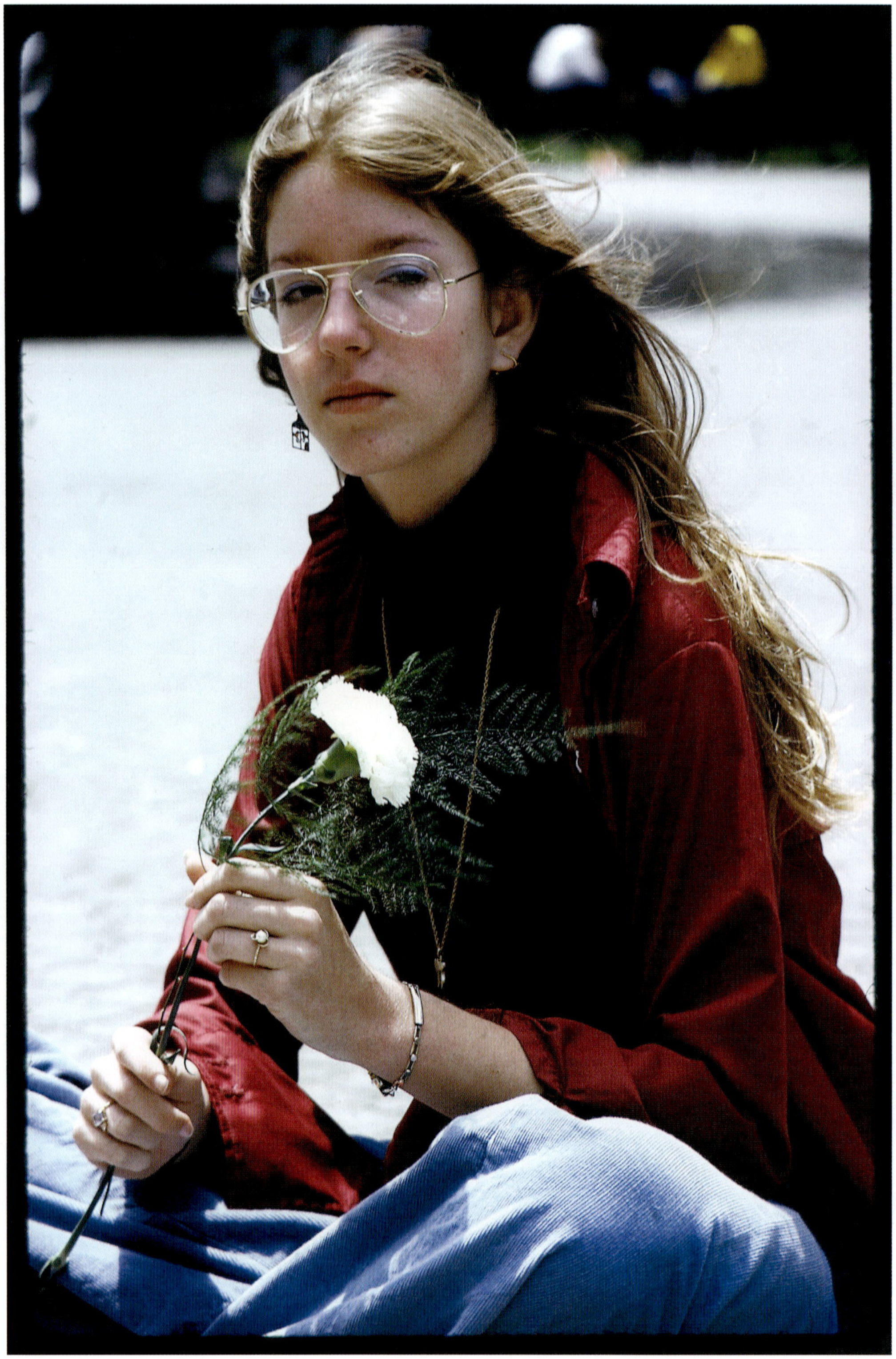

"Fountain Square in downtown Cincinnati is a public square that works for the city and its people in a myriad of ways: a flower vender has made a sale."

Tom Hubbard, Cincinnati, Ohio, May 1973

(412-DA-10706)

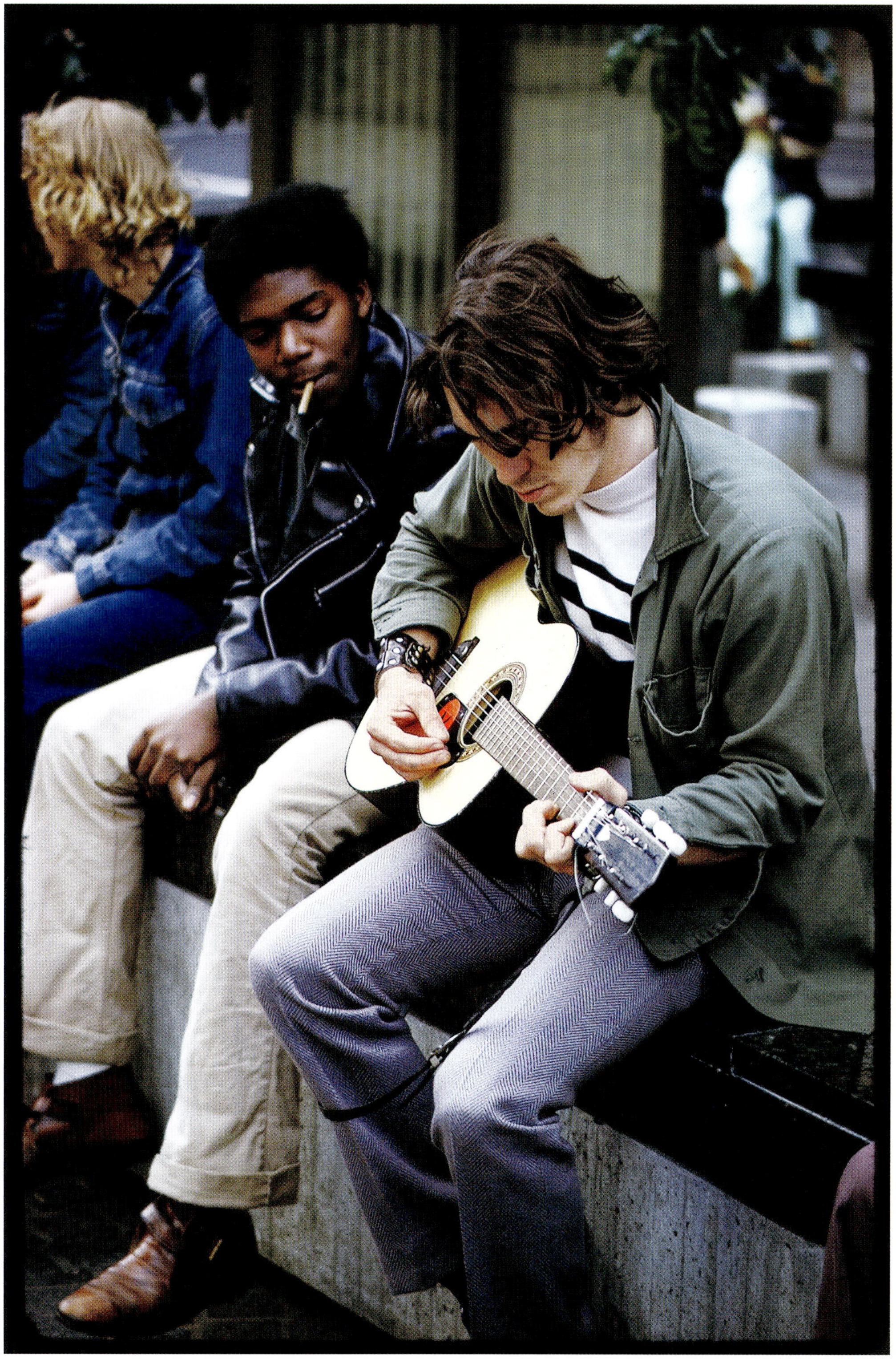

"Fountain Square in downtown Cincinnati is a public square that works for the city and its people in a myriad of ways: Saturday afternoon."

Tom Hubbard, Cincinnati, Ohio, May 1973

(412-DA-10693)

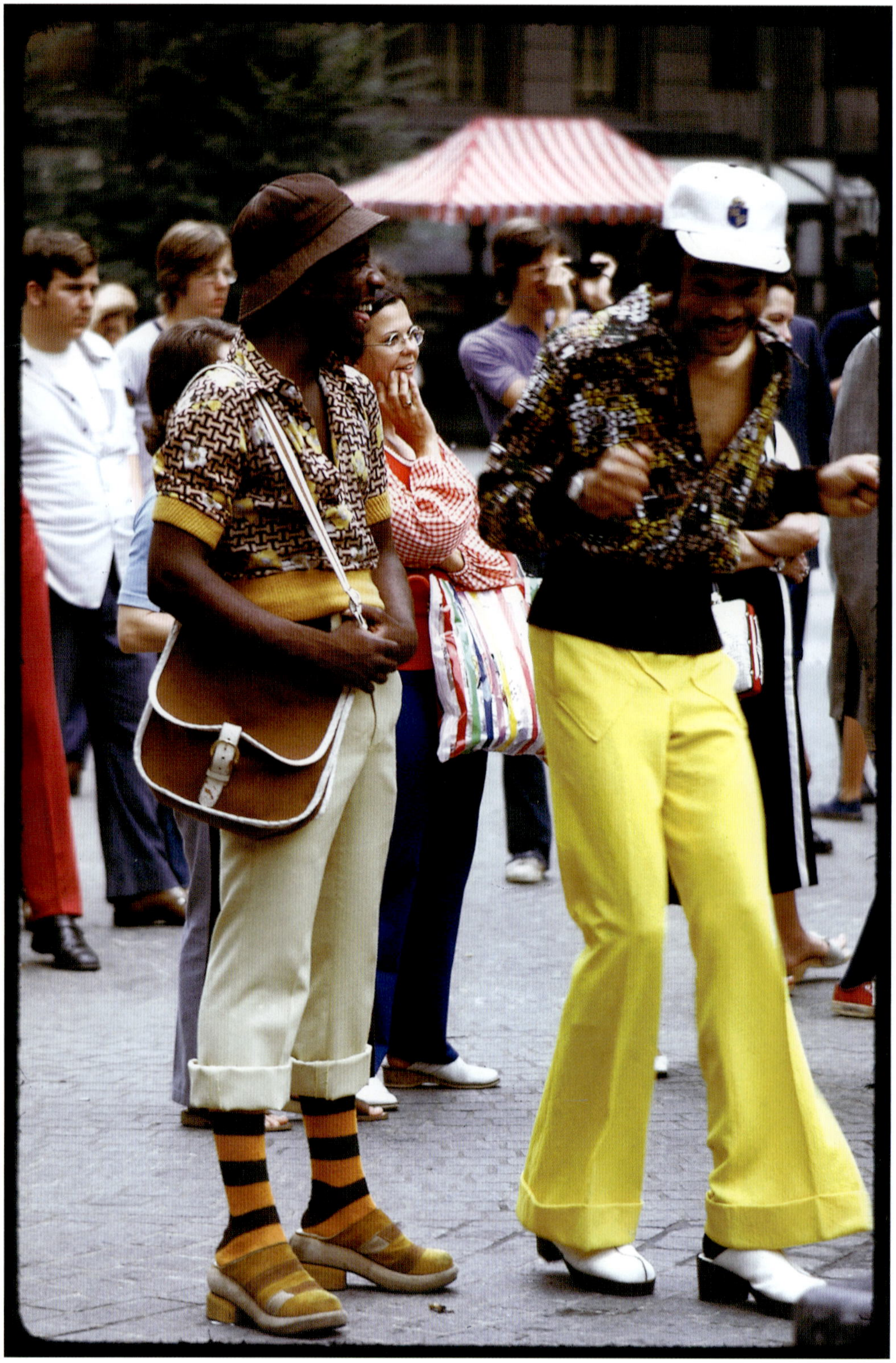

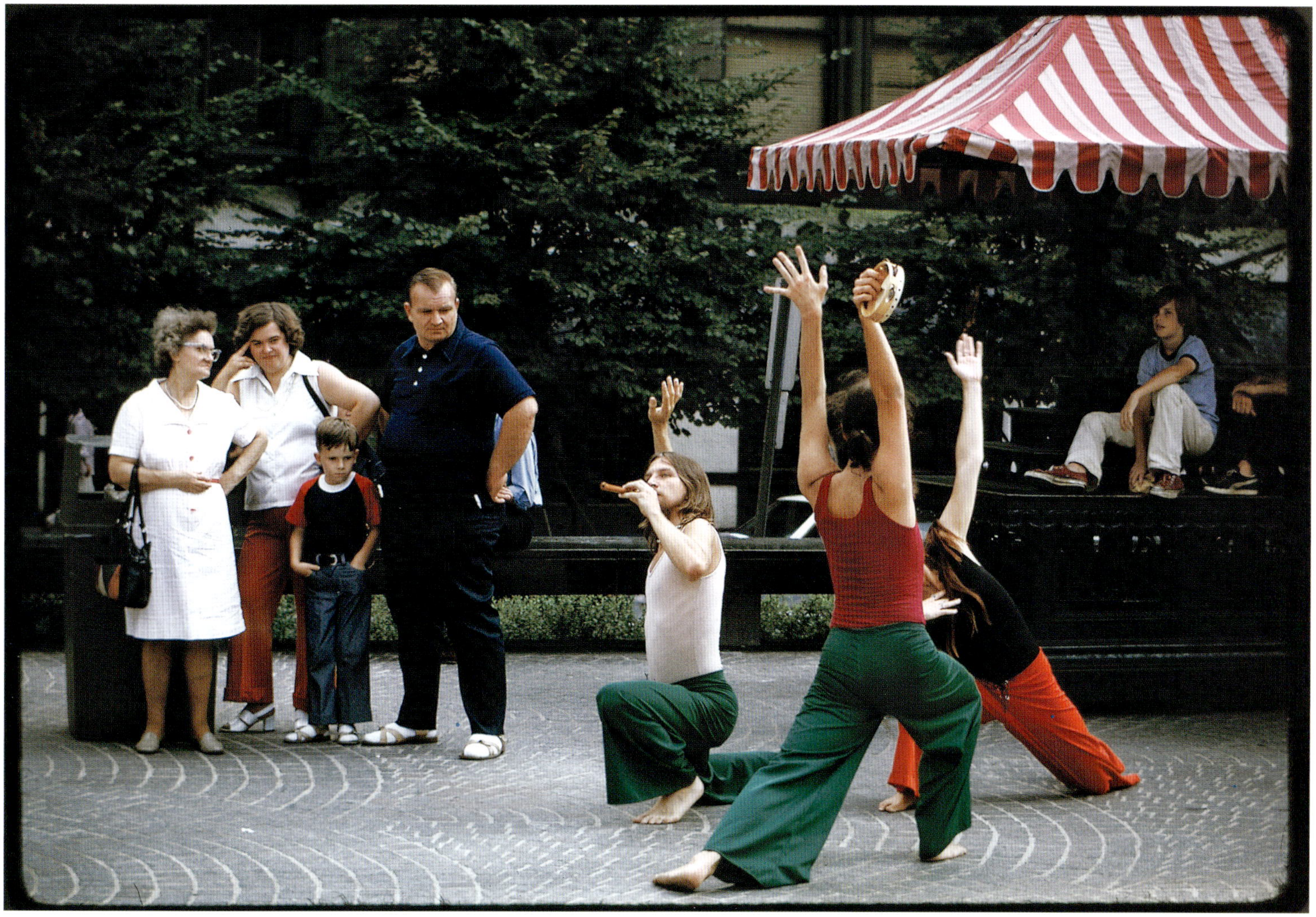

"'D'aug Days (pronounced dog) is a month long presentation of all the arts at downtown Cincinnati's immensely popular public plaza, Fountain Square. Dancers from New Media Theater, a Cincinnati group."

Tom Hubbard, Cincinnati, Ohio, August 1973

(412-DA-10824)

Opposite: "Far out style setters groove to music of Fountain Square."

Tom Hubbard, Cincinnati, Ohio, June 1973

(412-DA-10751)

DOCUMERICA Returns:

EPA's "State of the Environment"

EPA has created a 21st-century version of DOCUMERICA called "State of the Environment" to become a second time capsule of environmental progress and change. Since the project's launch on Earth Day 2011, the public has been contributing current digital photographs of life and the environment in the United States and around the world on Flickr: *http://bit.ly/sotephotos*.

Featured photos and project updates can be followed on the project blog: *www.epa.gov/stateoftheenvironment*. Participation is welcome through the end of 2013.

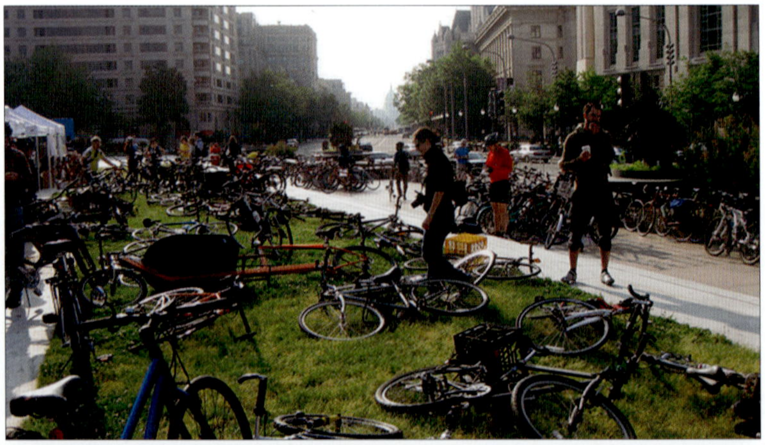

"Bike to Work Day," Washington, DC, May 2011
Mike Messner
Courtesy of the Environmental Protection Agency, State of the Environment

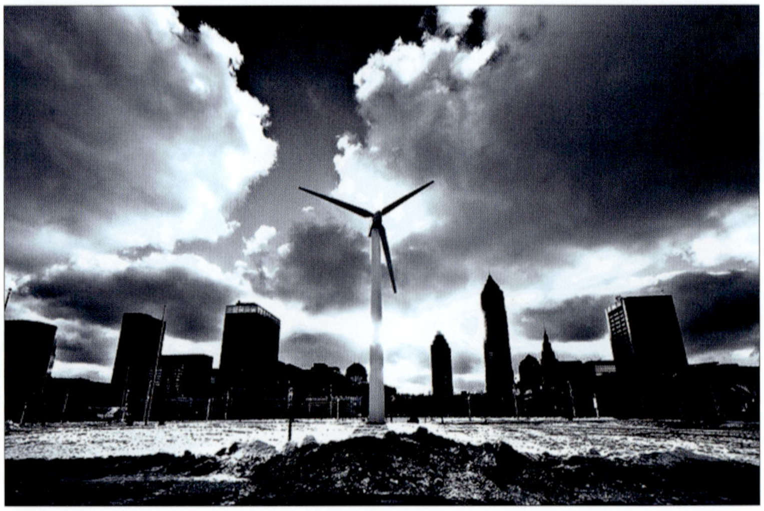

"Downtown Cleveland Landscape," Cleveland, Ohio, December 2010
Tate Davidson
Courtesy of the Environmental Protection Agency, State of the Environment

Further Reading

Bailey, Beth, and David Farber, eds. *America in the Seventies*. Lawrence: University of Kansas Press, 2004.

Berkowitz, Edward D. *Something Happened: A Political and Cultural Overview of the Seventies*. New York: Columbia University Press, 2006.

Borstelmann, Thomas. *The 1970s: A New Global History from Civil Rights to Economic Inequality*. Princeton: Princeton University Press, 2012.

Carroll, Peter N. *It Seemed Like Nothing Happened: America in the 1970s*. New York: Holt, Rinehart, Winston, 1982.

Cowie, Jefferson. *Stayin' Alive: The 1970s and the Last Days of the Working Class*. New York: The New Press, 2010.

Frum, David. *How We Got Here: The 70s: The Decade that Brought You Modern Life (For Better or Worse)*. New York: Basic Books, 2000.

Patterson, James T. *Grand Expectations: The United States, 1945–1974*. New York: Oxford University Press, 1996.

Schulman, Bruce J. *The Seventies: The Great Shift in American Culture, Society, and Politics*. Boston: De Capo, 2002.

Shubinski, Barbara Lynn. "From FSA to EPA: Project DOCUMERICA, the Dust Bowl Legacy, and the Quest to Photograph 1970s America," Ph.D. dissertation, The University of Iowa, 2009.

Zaretsky, Natasha. *No Direction Home: The American Family and the Fear of National Decline, 1968–1980*. Chapel Hill: University of North Carolina Press, 2007.

Notes

[1] Books that challenge our stereotypes of the 1970s include Peter N. Carroll, *It Seemed Like Nothing Happened: The Tragedy and Promise of America in the 1970s* (New York: Holt, Reinhart, and Winston,1982); Bruce J. Schulman, *The Seventies: The Great Shift in American Culture, Society, and Politics* (Boston: De Capo, 2002); David Frum, *How We Got Here: The 70s: The Decade that Brought You Modern Life (For Better or Worse)* (New York: Basic Books, 2000); Edward D. Berkowitz, *Something Happened: A Political and Cultural Overview of the Seventies* (New York: Columbia University Press, 2006); Beth Bailey and David Farber, eds., *America in the Seventies* (Lawrence: University of Kansas Press, 2004); Natasha Zaretsky, *No Direction Home: The American Family and the Fear of National Decline,* 1968–1980 (Chapel Hill, University of North Carolina Press, 2007); Jefferson Cowie, *Stayin' Alive: The 1970s and the Last Days of the Working Class* (New York: New Press, 2010); and Thomas Borstelmann, *The 1970s: A New Global History from Civil Rights to Economic Inequality* (Princeton: Princeton University Press, 2012).

[2] Gifford Hampshire, untranscribed video interview with Bruce Bustard, Arlington, Virginia, November 16, 1998. Item numbers 64.67 and 64.68; Records of the National Archives and Records Administration, Record Group (RG) 64, hereinafter cited as Hampshire/Bustard interview; Ken Light, *Witness in Our Time: Working Lives of Documentary Photographers* (Washington, DC: Smithsonian Institution Press, 2000); Barbara Lynn Shubinski, "From FSA to EPA: Project Documerica, the Dustbowl Legacy, and the Quest to Photograph 1970s America," (Ph.D. dissertation, The University of Iowa, 2009), pp.12–13.

[3] Hampshire/Bustard interview; Light, *Witness in Our Time*, pp. 165–67.

[4] Adam Rome, "'Give Earth a Chance': The Environmental Movement and the Sixties," *Journal of American History* (September 1990), pp. 525–54.

[5] On Richard Nixon's attitude toward the environmental movement, see James T. Patterson, *Grand Expectations: The United States 1945–1974* (New York: Oxford University Press,1996), p. 727; Dennis C. Williams, *The Guardian: EPA's Formative Years* (Washington, DC: U.S. Environmental Protection Agency,1993), p. 19; William Ruckelshaus oral history interview with Michael Gorn, January 1993, EPA web site at: *http://epa.gov/aboutepa/history/publications/print/ruck.html#bio* (accessed August 10, 2012); and Schulman, *The Seventies*, pp. 30–32.

[6] Hampshire/Bustard interview; Light, *Witness in Our Time*, p. 167; *Environmental News*, February 1972; General Correspondence, 1972–1973 folder, box 1, series 412-M, Records of the Environmental Protection Agency, RG 412, National Archives and Records Administration (DOCUMERICA records hereinafter cited as folder title, 412-series designation, NA); "Guidelines for Photographers," January 1972, in Lyon folder, assignment 0012, 412-P, NA.

[7] "Guidelines for Photographers." Commoner was one of several popular late-1960s and 1970s authors, sometimes referred to collectively as "The Prophets of Doom," who wrote books predicting worldwide environmental catastrophes of one sort or another. Commoner's most influential book, *The Closing Circle*, was published in 1971. In early 1970, he was on the cover of *Time*, which described him as "the Paul Revere of ecology." See *Time*, February 2, 1970.

[8] Hampshire/Bustard interview; James R. Gaines, "Documerica: Photographic Promises to Keep," *Saturday Review*, February 26, 1972. Gifford Hampshire to Ken Heyman, July 17, 1973, in Heyman folder, box 4, assignment 0035, 412-P, NA; on Hampshire urging photographers to read Hurley, see Hampshire to Heyman, July 17, 1973, cited above; Hampshire to Wil Blanche, January 18, 1973, Blanche folder, assignment 0044, 412-P, NA, and Hampshire to Arthur Greenberg, April 4, 1973, in Greenberg folder, box 7, assignment 0082, 412-P, NA. On Hampshire wanting to avoid "propaganda," see Light, *Witness in Our Time*, p. 170.

[9] DOCUMERICA Task Order for Gene Daniels, May 3, 1972, in Daniels folder, assignment 0023, 412-P, NA; DOCUMERICA Task Order for Frank Aleksandrowicz, March 22, 1973, Aleksandrowicz folder, assignment 0085, 412-P, NA.

[10] Yoichi Okamoto to Gifford Hampshire, April 1, 1974, Okamoto folder, assignment 0136, 412-P, NA.

[11] Shubinski, "From FSA to EPA," pp. 21–22.

[12] Ibid.

[13] While DOCUMERICA's photographic coverage made an effort to include a variety of Americans, only a few DOCUMERICA photographers, notably Okamoto, Woodson, and White, were minorities.

[14] Carroll, *It Seemed Like Nothing Happened*. Carroll intended that his title be taken ironically.

Index

Page numbers in italics refer to photographs.

NATIONAL ARCHIVES LOCATIONS

www.archives.gov/locations/

National Archives and Records Administration
Washington, DC, and College Park, Maryland
inquire@nara.gov

National Archives at Anchorage
alaska.archives@nara.gov

National Archives at Atlanta
atlanta.archives@nara.gov

National Archives at Boston
waltham.archives@nara.gov

National Archives at Chicago
chicago.archives@nara.gov

National Archives at Denver
denver.archives@nara.gov

National Archives at Fort Worth
ftworth.archives@nara.gov

National Archives at Kansas City
kansascity.archives@nara.gov

National Archives at New York City
newyork.archives@nara.gov

National Archives at Philadelphia
philadelphia.archives@nara.gov

National Archives at Riverside
laguna.archives@nara.gov

National Archives at San Francisco
sanbruno.archives@nara.gov

National Archives at Seattle
seattle.archives@nara.gov

National Archives at St. Louis and **National Personnel Records Center**
St. Louis, Missouri
mpr.center@nara.gov

PRESIDENTIAL LIBRARIES

Herbert Hoover Presidential Library and Museum
West Branch, Iowa

Franklin D. Roosevelt Presidential Library and Museum
Hyde Park, New York

Harry S. Truman Presidential Library and Museum
Independence, Missouri

Dwight D. Eisenhower Presidential Library and Museum
Abilene, Kansas

John F. Kennedy Presidential Library and Museum
Boston, Massachusetts

Lyndon Baines Johnson Presidential Library and Museum
Austin, Texas

Richard Nixon Presidential Library and Museum
Yorba Linda, California

Gerald R. Ford Presidential Museum
Grand Rapids, Michigan

Gerald R. Ford Presidential Library
Ann Arbor, Michigan

Jimmy Carter Presidential Library and Museum
Atlanta, Georgia

Ronald Reagan Presidential Library and Museum
Simi Valley, California

George Bush Presidential Library and Museum
College Station, Texas

William J. Clinton Presidential Library and Museum
Little Rock, Arkansas

George W. Bush Presidential Library and Museum
Lewisville, Texas